Rolling Sto

COA

STOCK

Second Edition

Colin J. Marsden

LONDON

IAN ALLAN LTD

First published 1987

ISBN 0 7110 1674 7

Published by Ian Allan Ltd, Shepperton, Surrey; and printed by Ian Allan Printing Ltd at their works at Coombelands in Runnymede, England

Below:
Although today the number of Mk 1 coaches has considerably diminished, a number of secondary services are still Mk 1- formed. A Cardiff-Bristol service headed by Class 33 No 33103 approaches Horfield on 8 June 1982 formed of Mk 1, CK, SK and TSO stock with a BG formed as the fourth vehicle.
Colin J. Marsden

Introduction

Probably the most frequently seen item of railway rolling stock is the passenger coach, and the non-passenger carrying coaching stock (NPCCS) vehicle of which there are 4,433 currently in service. Today the interest in non-traction stock is steadily growing, with many enthusiasts now collecting coach, NPCCS and even wagon numbers; thus books covering these subjects are now essential reading for the modern railway enthusiast. However, although there is now a wide interest in this subject, many people cannot identify the various types, such as FK, FO or BCK, and are not clear whether, for example, a vehicle is a Mk 2a, b, c, d, e or f. Other recognition complications can also exist in areas of braking and the systems of train heating.

This title in the *Rolling Stock Recognition* series has been prepared to assist in recognising the various vehicle types, and to show pictorially the most significant detail differences. Regrettably, for reasons of space, it may not be possible to show all minor detail differences, because within some fleets as many as 25 vehicle differences can be identified.

The stock included in this title is all types of passenger and NPCCS vehicles in traffic during September 1986. The main passenger fleet can be divided into three main categories — Mk 1, Mk 2 and Mk 3. The Mk 1 being the standard BR-designed coach introduced from 1951, the Mk 2 and subsequent sub-classes were introduced from 1964 with the final derivatives emerging in 1975. With the railway's continuing policy on modernisation the Mk 3 23m-long stock emerged from 1975, with the Mk 3b derivatives now in service.

The NPCCS fleet, slowly diminishing in size, is to be found in this title after the passenger stock, and gives vehicles under their TOPS identification system.

In recent years a number of revisions to passenger coaching stock and NPCCS number systems have been made. This has vacated many number series that have been used twice, especially where locomotives have been similarly numbered. This reorganisation of numbers has also improved provision for the grouping of vehicles with similar equipment — this is especially prevalent in the NPCCS field.

Since the first impression of this title in 1983, the single largest change in coaching stock has been livery. From 1984 the InterCity sector changed its colours from blue/grey to two-tone grey with a white and red waist-height strip; this scheme is now progressively being applied as stock passes through main works for overhaul. ScotRail decided to be different on the livery theme for its internal stock, which has emerged in the new InterCity sector colours but with a blue waist band in place of the red. In May 1986 a further livery advance was made when *Network SouthEast* was launched, leading to coaches in blue, red and white livery. A number of vehicles that are used by the provincial sector may well retain their present scheme. Most NPCCS vehicles still carry the blue/grey colours with the exception of the remaining company-built stock, some NCVs, and all NJ, NK, NL, NP and NX types which are in all-over rail blue colour.

It is hoped that this revised edition of *Rolling Stock Recognition 1 — Coaching Stock* will give readers some assistance in the understanding and identification of the large variety of CS and NPCCS vehicles seen on the BR network today.

The Editor would be pleased to receive any information regarding significant features of rolling stock that have not been included. These should be sent, together with an illustration if possible, to The Editor, *Rolling Stock Recognition*, c/o Ian Allan Ltd, Coombelands House, Coombelands Lane, Addlestone, Weybridge, Surrey KT15 1HY.

Worcester Park, Surrey
September 1986

Colin J. Marsden

Cover:
A Network SouthEast liveried Mk 1 and Mk 2 stand inside Old Oak Common depot.
Brian Morrison

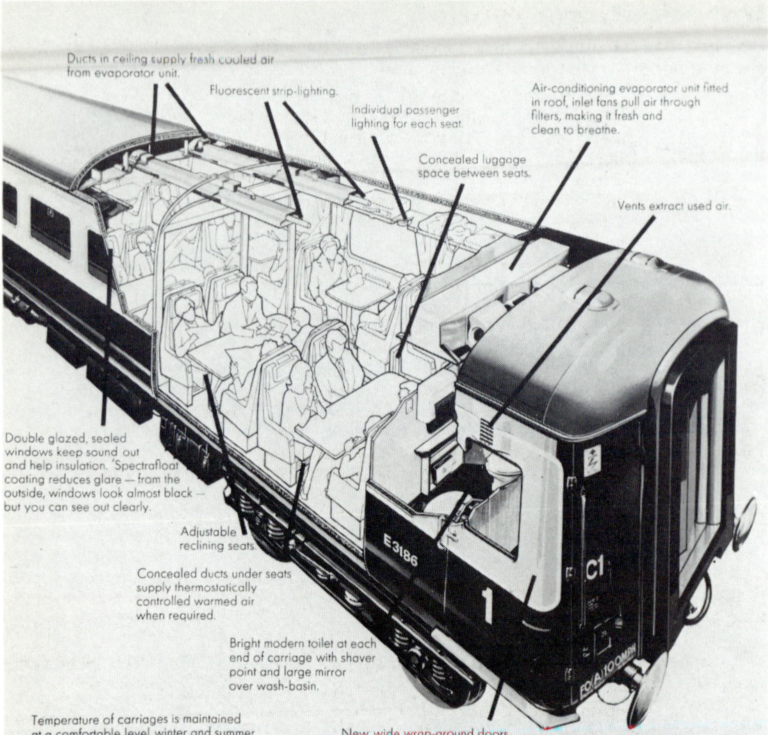

THE MARK IID CARRIAGE.

Vehicle Bogie Types

A number of different bogie types are to be found under BR coaching stock. These are identified by a letter/number code system.

B1	(Standard) Double-bolster bogie — fitted with leaf springs. Normally the speed is restricted to 90mph, but special maintenance on some gives a speed of 100mph. Weight 6.1 tons
B1	(Heavy Duty) Basically the same as above, but with a weight of 6.5 tons
B2	Final design of B1 bogie fitted under some van stock only. Weight 6.2 tons
B4	Steel fabricated bogie fitted with coil springs, and restricted to 100mph or with special maintenance 110mph. Weight 5.2 tons
B5	Heavy duty version of the B4, maximum speed 100mph. Weight 5.3 tons
BT10	New generation fabricated bogie, fitted with air suspension and disc brakes. Maximum speed 125mph
BT15	Similar to BT10, but with rubber primary suspension
BT7	This type of bogie fitted under a few Mk 3 vehicles is a derivative of the BX1 type as fitted to modern EMU stock. Fitted with air suspension and disc brakes
CW	(Commonwealth) Heavy cast-steel bogie with coil springs, designed to improve ride of previous B1 and B4 types. Maximum speed 100mph. Weight 6.8 tons
Single Bolster	1950s design leaf spring bogie — fitted to a small number of SKs only. Maximum speed 90mph. Weight 5.3 tons

Above left:
Section view of a Mk 2d FO coach, detailing the equipment in the Mk 2 air-conditioned range. The seating details only apply to first class vehicles. *BR*

Left:
The **B1** standard double-bolster leaf-spring bogie. Usually vehicles carried on these are restricted to 90mph, but with special maintenance 100mph operation is permitted.
Colin J. Marsden

Above:
CW Bogie: To improve riding qualities the Commonwealth range was introduced, being a heavy cast-steel assembly with coil springs; maximum speed 100mph.
Colin J. Marsden

Coach Classification

Passenger

BCK — Brake Composite Corridor
BFK — Brake First Corridor
BFO — Brake First Open
BSK — Brake Second Corridor
BSO — Brake Second Open
BSO(T) — Brake Second Open with Micro Buffet
CK — Composite Corridor
DBSO — Driving Brake Second Open
FK — First Corridor
FO — First Open
RBR — Restaurant Buffet Refurbished
RKB — Restaurant Kitchen Buffet
RMB — Restaurant Miniature Buffet
RFB — Restaurant First Buffet
RSM — Restaurant Second Modular
SK — Second Corridor
SLE — Sleeper Either Class
SLEP — Sleeper Either Class with Pantry
SO — Second Open
TSO — Tourist Second Open
TSO(T) — Tourist Second Open with Micro Buffet
PFB — Pullman First Brake
PFK — Pullman First Kitchen
PFP — Pullman First Parlour

Vale of Rheidol Railway Stock

BCO — Brake Composite Open
BY — Full Brake
SO — Second Open

Non-passenger Carrying Coaching Stock

B	NF	— Bogie Brake
BG	NB, NC, ND, NE, NH	— Brake Gangway
BFB	NW	— Bogie Flat Bullion
CCT	NO, NP	— Covered Carriage Truck
FVV	NG	— 10-ton Carflat
GUV	NJ, NK, NL, NY	— General Utility Vehicle
PMV	NQ	— Parcels and Miscellaneous Van
POS	NS	— Post Office Sorting
POT	NT	— Post Office Tender
POT(B)	NU	— Post Office Tender with Brake
PVG	NL	— Packing Van Gangwayed
SLB	NW	— Security Bullion Vehicle
TCV	NV	— Two-tier Car-Carrying Vehicle

TOPS Passenger Stock Codes

1 Letter codes

AA	Gangwayed Corridor
AB	Gangwayed Corridor Brake
AC	Gangwayed Open (2+2 layout)
AD	Gangwayed Open (2+1 layout)
AE	Gangwayed Open Brake
AF	Gangwayed Driving Open Brake
AH	Kitchen Buffet
AJ	Restaurant Buffet (with kitchen)
AK	Kitchen
AM	Restaurant Open
AN	Miniature Buffet
AO	Private Owner Vehicles (Passenger)
AS	Sleeping Stock
AT	Royal Train Stock
AU	Sleeping Stock
AX	Narrow Gauge Stock (V of R Railway)
AY	Narrow Gauge Stock (V of R Railway)

2 Class codes (Seating)

1	First
2	Second
3	Composite
4	Unclassified
5	None

3 Suffix letters

O	Pre-nationalisation
1	Mk 1
Z	Mk 2
A	Mk 2a
B	Mk 2b
C	Mk 2c
D	Mk 2d
E	Mk 2e
F	Mk 2f
G	Mk 3a
H	Mk 3b

Above:
When the Royal Train is in use it is normal practice for a minimum of four vehicles to be used. The train illustrated, headed by SR 'Royal Locomotive' No 73142 *Broadlands*, is formed of five vehicles, Nos 2906, 2901, 2900, 2902, 2905. The train was photographed approaching New Malden on 4 August 1983 when HM The Queen travelled from Waterloo to Southampton Docks. *Colin J. Marsden*

Type: **BCK** TOPS Code: **AB31**

Mk: 1
Number range: 21238-21275
Introduced: 1961-64
Built by: BR Swindon/Derby
Regions of allocation: ScR/LMR/WR/SR
Seats: 12 first, 18 second or 12 first, 24 second
Speed (max): 100mph (161km/h)
Total number in fleet: 16
Brake type: Vacuum/air
Heating type: Dual/electric
ETH index: 3
Bogie type: CW
Length: 64ft 6in (19.65m)
Height: 12ft 9½in (3.89m)
Width: 9ft 3in (2.81m)
Weight: 36 tonnes

Above:
Mk 1 BCK: To provide short brake and guards facilities in CK coaches a fleet of BCK (Brake Composite Corridor) vehicles were constructed in the period 1953-64, by various BR works and private companies. Looking from the saloon end, corridor side, No 21268 is illustrated, this being an air-braked, electrically heated vehicle. *Colin J. Marsden*

Above:
Mk 1 BCK: The seating arrangement of these vehicles is of two basic layouts: 12 first/18 second, and 12 first/24 second, this being dependent on whether the compartments have arm-rests or not. Initially cars built for ER/LMR/NE and ScR operation were designed for six persons per compartment, while those for the WR and SR were laid out for eight persons. Car No 21268 is seen from the brake end, the two first class compartments being adjacent to the guard's compartment. *Colin J. Marsden*

Above:
Mk 1 BCK: The present fleet of 16 BCKs were built in the early 1960s by BR at Swindon and Derby. The recognition between BCK, BFK and BSK vehicles is comparatively easy if it is remembered that on BCKs the first class band at cantrail height is in the middle of the coach, and that on BFK and BSK the luggage van is much larger. A train formed of a BCK and TSO headed by Class 73 No 73006 approaches Hampton Court Junction on 24 January 1981. *Colin J. Marsden*

Type: **BFK** TOPS Code: **AB1**

Mk:	1	2
Number range:	17003-17026	17028-17055
Former number range:	14003-14026	14028-14055
Introduced:	1959-63	1966
Built by:	BR Ashford/Swindon	BR Derby
Regions of allocation:	ScR/WR/LMR	ScR/WR/ER/LMR
Seats:	24 first	24 first
Speed (max):	90-100mph (145-161km/h)	100mph (161km/h)
Total number in fleet:	13	24
Brake type:	Vacuum	Vacuum
Heating type:	Dual	Dual
ETH index:	2	4
Bogie type:	B1/B4/CW	B4
Length:	64ft 6in (19.65m)	66ft 0in (20.11m)
Height:	12ft 9½in (3.89m)	12ft 9½in (3.89m)
Width:	9ft 3in (2.81m)	9ft 3in (2.81m)
Weight:	34 tonnes	32 tonnes
Notes:		

Above:

Mk 1 BFK: To provide brake accommodation in FK (First Corridor) coaches a fleet of BFK (Brake First Corridor) vehicles was constructed, accommodation being in four compartments, with a toilet at the outer end. Vehicle No 14025 is viewed from the accommodation end, compartment side. This vehicle, mounted on CW bogies, bears the Motorail legend on the end, indicating that at one time the vehicle was restricted to this use. *Colin J. Marsden*

2a	2b	2c	2d
17056-17103	17107	17113-17138	17139-17172
14056-14103	14107	14113-14138	14139-14172
1967-68	1969	1969-70	1971-72
BR Derby	BR Derby	BR Derby	BREL Derby
ScR/WR/ER	WR	ScR/LMR/WR/ER	ScR/LMR/ER
24 first	24 first	24 first	24 first
100mph (161km/h)	100mph (161km/h)	100mph (161km/h)	100mph (161km/h)
40	1	25	34
Air/Vacuum*	Air	Air	Air
Dual	Dual	Dual	Electric
4	4	4	5
B4	B4	B4	B4
66ft 0in (20.11m)	66ft 0in (20.11m)	66ft 0in (20.11m)	66ft 0in (20.11m)
12ft 9½in (3.89m)	12ft 9½in (3.89m)	12ft 9½in (3.89m)	12ft 9½in (3.89m)
9ft 3in (2.81m)	9ft 3in (2.81m)	9ft 3in (2.81m)	9ft 3in (2.81m)
32 tonnes	32 tonnes	32 tonnes	34 tonnes
*All vehicles built air-braked but 19 modified for vacuum operation. Some vehicles now reclassified as BSK and renumbered			Vehicles 17162-72 require 800V train heat supply

Above:
Mk 1 BFK: The present fleet of Mk 1 BFKs numbers only 13, being allocated to the ScR, WR, LMR and ER. All vehicles are painted in blue/grey livery except No 17023 which is in InterCity grey/red. No 17003, the earliest surviving vehicle, is illustrated bearing the ScotRail legend. *Michael J. Collins*

Top:
Mk 2 BFK: The BFK design was carried on in the Mk 2 builds. In 1966, 28 Mk 2s were built, to the same internal layout as the Mk 1s. All vehicles have 24 seats and one toilet, and are mounted on B4 bogies. No 17053 is shown from the brake end which has a 1½-tonne luggage capacity. *Michael J. Collins*

Above:
Mk 2a BFK: The Mk 2a vehicles are almost identical to the Mk 2 build, with doors and windows being in the same positions. Internally each compartment has an aluminium sliding door, compared to a wooden one on previous fleets. External end doors are also revised being of the folding type coloured green. Car No 14072, now renumbered 17072, is illustrated. *Colin J. Marsden*

Top:
Mk 2a BFK: When built, all Mk 2a BFK vehicles were fitted with air brake equipment; however, over the years 19 were converted to vacuum braking. Mk 2a No 17076, an air-braked vehicle allocated to LA, is shown from the saloon end compartment side. Underframe equipment consists of heat, light and brake modules. *Colin J. Marsden*

Above:
Mk 2b BFK: Only four Mk 2b BFK vehicles were built and only one, No 17107, remains in traffic. Accommodation is again for 24 in four compartments. The luggage van can carry up to 3 tonnes of luggage. Withdrawn No 17110 is illustrated. *Colin J. Marsden*

Top:

Mk 2c BFK: In 1971/72 a fleet of 34 Mk 2c BFKs emerged following the same basic design as the Mk 2b except that internally lower ceilings and ducting for air conditioning was installed, but no air conditioning modules were fitted. Mk 2c vehicles can be recognised against the Mk 2b versions as the latter only having three or four roof-mounted ventilators compared to 12 on the Mk 2b. No 17120 is illustrated from the brake end; underframe equipment visible consists of train lighting batteries and brake units. *Colin J. Marsden*

Above:

Mk 2d BFK: A fleet of 34 Mk 2d BFKs were built in 1971/72, all fitted with air braking and air conditioning, and are thus immediately recognisable. Although the distinction between these and the Mk 2d BSOs can be difficult to establish, the best way is to refer to the yellow first class stripe on the BFK or, of course, the number. Car No 17156 is illustrated. *Colin J. Marsden*

The **BFK** fleet can be found operating on all regions providing brake accommodation for similar train formations. This interesting shot shows a Mk 2c BFK forming the brake accommodation for a track recording train at Westbury on 30 November 1983.
Colin J. Marsden

Type: **BFO** TOPS Code: **AE1**

Mk: 3b
Number range: 17173-17175
Introduced: 1985-86
Built by: BREL Derby
Regions of allocation: LMR
Seats: 36 first
Speed (max): 110mph (177km/h)
Total number in fleet: 3
Brake type: Air
Heating type: Electric
ETH index: 5
Bogie type: BT10
Length: 75ft 0in (22.86m)
Height: 12ft 9in (3.82m)
Width: 8ft 11in (2.71m)
Weight: 34 tonnes
Notes: Fitted with hydraulic handbrake

Above:
Mk 3b BFO. From early 1986 a new coach designation appeared — Brake First Open (BFO). Three vehicles of this type have been built for WCML services, each coach seats 36 first class passengers in the open 2+1 layout, with a small guard's office and luggage compartment at one end with a capacity of 0.75 tonnes of luggage. This detail illustration of car No 17173 shows the guard's area. The main guard's door is of the outward-opening type, with a small secondary hinged door to increase the entrance width. The bogie is of the BT10 type. *C. J. Tuffs*

Right:
Mk 3b BFO. General view of BFO from guard's end. Recognition of the type should cause no problem as no other comparable vehicle in the locomotive-hauled fleet exists, the closest vehicle type being the TGS in the IC125 fleet, but these have a smaller guard's area. There is one toilet at the passenger end. *C. J. Tuffs*

17175
C3

Type: **BSK** TOPS Code: **AB2**

Mk:	1	2a
Number range:	34521-35486	35500-35503
Former number range:	—	170XX (BFK)
Introduced:	1955-63	Orig: 1967-68. As BSK: 1986
Built by:	Metro-Cammell/Roberts/ Gloucester RC&W/BR Wolverton	BR Derby
Regions of allocation:	ScR/LMR/WR/SR/ER	ER
Seats:	24 or 32 second (dependent on build)	24 second
Speed (max):	90-100mph (145-161km/h)	100mph (161km/h)
Total number in fleet:	217	4
Brake type:	Vacuum/air	Vacuum
Heating type:	Dual/electric	Electric
ETH index:	2	4
Bogie type:	B1/B4/CW	B4
Length:	64ft 6in (19.65m)	66ft 0in (20.11m)
Height:	12ft 9½in (3.89m)	12ft 9½in (3.89m)
Width:	9ft 3in (2.81m)	9ft 3in (2.81m)
Weight:	34 tonnes	32 tonnes
Notes:	Some SR allocated vehicles have 750V train heating	

Above:

Mk 1 BSK. Under the 1951 coaching stock plan, the brake vehicle containing second class passenger accommodation was classified BSK (Brake Second Corridor). The coach basically consists of a brake van and four passenger compartments, with a toilet at the passenger end. Initially vehicles were fitted with vacuum brakes and steam heating but over the years air brakes and electric heating have been installed. BSK No 34641 is illustrated from the brake van end and compartment side, the brake van being able to accommodate 2½ tonnes of luggage. *Colin J. Marsden*

Below:
Mk 1 BSK. Two different internal configurations of BSK exist, those set out for 24 persons and those for 32, this being dependent on whether armrests are provided or not. All vehicles have four compartments fed by a side corridor. Vehicle No 34940 illustrated is mounted on B1 bogies and is one of a batch allocated to Clapham Junction and fitted with 750V train heating. *Colin J. Marsden.*

Below:
Mk 1 BSK. BSKs numbered above 35274 have metal window frames, melamine interior panels, and run on B4 or CW bogies. Two vehicles, Nos 35454 and 35465, have been refurbished, given a modern internal decor, fluorescent lighting and painted in the new InterCity colours. Vehicle 35465 is illustrated. The BSK fleet are the only passenger-carrying vehicles equipped with two double-leaf doors each side of the brakevan, thus giving another recognition factor of the type. Some differences in stepboards can also be found — most only have boards under the doors but on a few examples the boards run the full length of the vehicle. *Colin J. Marsden*

Type: **BSO** TOPS Code: **AE2**

Mk:	1	2	2a
Number range:	9227, 9312, 9364	9382-9416	9417-9438
Introduced:	1955-56	1966	1967-68
Built by:	BR Doncaster/ Wolverton	BR Derby	BR Derby
Regions of allocation:	ScR	ScR/ER	ScR/LMR/ER
Seats:	39 second	31 second	31 second
Speed (max):	90mph (145km/h)	100mph (161km/h)	100mph (161km/h)
Total number in fleet:	3	22	20
Brake type:	Vacuum/dual	Vacuum	Air
Heating type:	Steam/dual	Dual	Dual
ETH index:	3	4	4
Bogie type:	B1/CW	B4	B4
Length:	64ft 6in (19.65m)	66ft 0in (20.11m)	66ft 0in (20.11m)
Height:	12ft 9½in (3.89m)	12ft 9½in (3.89m)	12ft 9½in (3.89m)
Width:	9ft 3in (2.81m)	9ft 3in (2.81m)	9ft 3in (2.81m)
Weight:	34 tonnes	31.5-33.5 tonnes	32 tonnes
Notes:		Some vehicles are wired for push-pull operation, and have disc brakes. Some vehicles rebuilt to BSO(T)	

Above:

Mk 1 BSO. Apart from the obvious difference of having an open layout, the main change between the Mk 1 BSK and BSO stock is that the BSO features five seating bays whereas on the former there are only four compartments. This view shows vehicle No 9350 (now withdrawn) from the brake end, looking along the coaches with the guard's office, passenger door, five seating bays, toilet and passenger door. Only three vehicles of this type remain in service. *B. J. Nicolle.*

2c	2d	2e	2f
9439-9478	9479-9495	9496-9509	9513-9539
1970-71	1971	1972	1974
BREL Derby	BREL Derby	BREL Derby	BREL Derby
ScR/LMR/WR/ER	LMR/ER	LMR	LMR
31 second	31 second	31 second	31 second
100mph (161km/h)	100mph (161km/h)	100mph (161km/h)	100mph (161km/h)
40	16	14	16
Air	Air	Air	Air
Dual	Electric	Electric	Electric
4	5	5	5
B4	B4	B4	B4
66ft 0in (20.11m)	66ft 0in (20.11m)	66ft 0in (20.11m)	66ft 0in (20.11m)
12ft 9½in (3.89m)	12ft 9½in (3.89m)	12ft 9½in (3.89m)	12ft 9½in (3.89m)
9ft 3in (2.81m)	9ft 3in (2.81m)	9ft 3in (2.81m)	9ft 3in (2.81m)
32 tonnes	33 tonnes	33 tonnes	33 tonnes

Above:

Mk 2 BSO. The fleet of 30 Mk 2 vehicles were built on the same basic shell as the BFKs being of standard Mk 2 design, with non wrap-around doors. On the Mk 2 and Mk 2a BSO builds, passenger accommodation conforms to first class spacing with wider tables than those found on second class stock. Mk 2 No 9410 is illustrated. *B. J. Nicolle*

Left:
Mk 2a BSO. When the Mk 2a BSO build was produced in 1967/68, the Mk 2a BFK body shell was utilised, and thus the internal layout is quite spacious. Externally the Mk 2 and Mk 2a vehicles are almost identical and recognition is best achieved by reference to vehicle numbers. All vehicles are allocated to the Midland, Scottish and Eastern Regions. No 9437 is illustrated from the saloon end, with the toilet area nearest the camera, after the entrance vestibule. *Michael J. Collins*

Below:
Mk 2c BSO. No Mk 2b BSO vehicles were built, but in 1970 a batch of 40 Mk 2c cars were constructed. This time the comparable BFK shell was not utilised and therefore the accommodation is in standard second class style. Wrap-around end doors were fitted and a smaller sized toilet window installed, providing good recognition factors. Car No 9465 is illustrated from the saloon end. *Colin J. Marsden*

Right:
Mk 2d BSO. In 1971 a fleet of 16 Mk 2d vehicles were constructed. Recognition against the previous builds is immediate as air conditioning is provided, giving no opening windows except in the doors. The guard's and luggage accommodation (nearest end in picture), occupies around half the vehicle length. Vehicle No 9494 is illustrated. The Mk 2e BSO fleet are identical to the Mk 2d build from the outward appearance, and recognition is best achieved by reference to the numbers. (Mk 2e type not illustrated). *Michael J. Collins*

Norwich
TIME NOW
12 28
Way
NORWICH UNION
Inter-City
C1
E9494
Air Bra

Above:
Mk 2f BSO. A fleet of 33 Mk 2f BSO vehicles were built in 1974 for WCML operation, however today only 17 remain in service as the balance have been rebuilt to DBSO stock. The external layout of cars is identical to the Mk 2e build but the interiors are styled with InterCity 70 seats. Mk 2f cars also have Temperature Ltd air conditioning equipment whereas the Mk 2es are fitted with Stones equipment. Car No 9521, painted in the latest InterCity colours, is illustrated from the saloon end. *Colin J. Marsden.*

Below:
BSO vehicles are allocated to all regions except the SR, and are usually used in formations of similar type stock to provide guard/luggage accommodation. A Mk 2c vehicle is formed as the second coach in the 09.50 Edinburgh-Plymouth of 22 June 1981, seen passing Dawlish Warren headed by Class 50 No 50036 *Victorious.*
Colin J. Marsden.

Type: **BSO(T)** TOPS Code: **AH2**

Mk:	1	2
Number range:	9000-9016	9100-9107
Former number range:	92XX, 93XX (random)	93XX, 94XX (random)
Introduced:	Orig: 1955-63. As BSO(T): 1971-72	Orig 1966. As BSO(T): 1986
Built by:	Orig: BR Wolverton/ Doncaster	Orig: BR Derby
Region of allocation:	ScR	ScR
Seats:	31 second	23 second
Speed (max):	90-100mph (145-161km/h)	100mph (161km/h)
Total number in fleet:	9	8
Brake type:	Vacuum/dual	Vacuum
Heating type:	Steam/dual	Dual
ETH index:	3	4
Bogie type:	B1/CW	B4
Length:	64ft 6in (19.65m)	66ft 0in (20.11m)
Height:	12ft 9½in (3.89m)	12ft 9½in (3.89m)
Width:	9ft 3in (2.81m)	9ft 3in (2.81m)
Weight:	34 tonnes	31.5 tonnes
Notes:	Some vehicles have tungsten lighting	

Below:
Mk 1 BSO(T). So as to provide a catering facility on certain routes where a full buffet car is not economical but where a buffet facility is still justified, it was decided to convert one seating bay (at the opposite end to the guard's van) to a service counter on some BSO vehicles, where hot and cold drinks and light snacks could be served. During conversion the passenger toilet was removed in favour of a steward's washing area. Vehicle No 9012 is illustrated with the buffet area at the far end. *Colin J. Marsden*

Type: **CK** — TOPS Code: **AA3**

Mk:	1	2c
Number range:	7074-7996	7550-7561
Former number range:	16074-16258, 15626-15996	13XXX fleet (FK)
Introduced:	1958-63	Orig: 1969-70; rebuilt (CK): 1985
Built by:	Metro-Cammell/BR Derby/ Wolverton	BR Derby
Regions of allocation:	ScR/LMR/WR/ER	ScR
Seats:	24 first/18 second, or 24 first/24 second	18 first/24 second
Speed max:	90-100mph (145-161km/h)	100mph (161km/h)
Total number in fleet:	91	5
Brake type:	Vacuum	Air
Heating type:	Steam/electric/dual	Dual
ETH index:	3 or 4	4
Bogie type:	B1/CW	B4
Length:	64ft 6in (19.65m)	66ft 0in (20.11m)
Height:	12ft 9½in (3.89m)	12ft 9½in (3.89m)
Width:	9ft 3in (2.81m)	9ft 3in (2.81m)
Weight:	34 tonnes	33 tonnes
Notes:		More vehicles are likely to be converted

Above:
Mk 1 CK. To replace many old pre-nationalisation vehicles a large number of BR designed compartment vehicles were introduced from 1951 and classified CK (Composite Corridor), first class seating being provided in four compartments and second class in three. Car No 15938 (now No 7938) is illustrated from the corridor side at the first class end. *Colin J. Marsden*

Above:
Mk 1 CK. Two distinct variants of CK were built, those intended for ER, LMR, NER and ScR operation, and those for SR and WR use. The first type seat six per second class compartment and the latter eight. CK No 7207 is mounted on CW bogies and is shown from the corridor side, second class end. *Michael J. Collins.*

Above:
Mk 1 CK. Along the corridor side the CK has four entrance doors, two in each of the first and second class areas; there is also a toilet at each end. The external configuration of the CK is unique and recognition should therefore cause no problems. Vehicle No 7973, a vacuum brake, steam heat example mounted on B1 bogies, is illustrated.
Michael J. Collins

Above:
Mk 1 CK. Today 91 CKs still remain in traffic, allocated to all regions except the SR. Vehicle No 7076 is illustrated. *Colin J. Marsden*

Below:
Mk 2c CK. In 1985 four Mk 2c FK vehicles were reclassified to CK, the second class area extending to four compartments and the first class to three. Internally the first class trims such as carpets and headrests have been removed. All five vehicles are allocated to Inverness and are usually deployed on Inverness-Glasgow duties. Car No 7551 (formerly FK 13551) is illustrated at Glasgow. *Keith Grafton*

A **CK** forms the first vehicle behind Class 47 No 47028 photographed passing Bourton near Swindon on 5 January 1980 with an up special from South Wales to Paddington.
Colin J. Marsden

Type: **DBSO** TOPS Code: **AF2**

Mk: 2f
Number range: 9701-9714
Former number range: From 95XX BSO fleet
Introduced: Orig: 1974; converted to DBSO: 1979-86
Built by: Orig: BREL Derby; converted: BREL Glasgow
Region of allocation: ScR
Seats: 32 second
Speed (max): 100mph (161km/h)

Total number in fleet: 13
Brake type: Air
Heating type: Electric
ETH index: 5
Bogie type: B4
Length: 66ft 0in (20.11m)
Height: 12ft 9½in (3.89m)
Width: 9ft 3in (2.81m)
Weight: 34 tonnes
Notes: All fitted with cowcatchers

Below:
Mk 2f DBSO. To facilitate driving positions at the remote end of the Edinburgh-Glasgow push-pull service a fleet of 10 DBSO (Driving Brake Second Open) vehicles were converted from Mk 2f BSO stock in 1979/80, a further four being converted in 1985/86. Part of the luggage van was converted into a driving compartment. On the front ends, windows were made in the two side and central positions, and marker lights, tail lights and headlights installed. DBSO No 9710, painted in ScotRail colours, is illustrated. *Colin J. Marsden*

Above right:
Mk 2f DBSO. The DBSO passenger accommodation remains the same as in BSO days, but the toilet is removed. All vehicles are mounted on B4 bogies fitted with air disc brakes, electric heating, public address and driver-guard communication equipment. All are painted in ScotRail colours. Car No 9705 is illustrated from the saloon end. *Keith Grafton*

Bottom right:
Following a fatal accident at Polmont in late 1984 when a DBSO-led train collided with a bullock at high speed and derailed, 'cowcatchers' were progressively fitted to all DBSO stock during 1985, upon recommendation by the inspecting officers. The 'cowcatcher' fitment is displayed here on car No 9705. Note this modification has meant the removal of the buck-eye coupling and revision to the air pipes. *Keith Grafton*

ScotRail

9705

Type: **FK** TOPS Code: **AA1**

Mk:	1	2
Number range:	13085-13344	13384-13427
Introduced:	1954-62	1964-65
Built by:	BR Swindon	BR Derby
Regions of allocation:	LMR/WR/ER	ScR/WR/SR/ER
Seats:	42 first	42 first
Speed (max):	100mph (161km/h)	100mph (161km/h)
Total number in fleet:	33	32
Brake type:	Vacuum/dual	Vacuum/air*
Heating type:	Dual	Dual
ETH index:	3	4
Bogie type:	B1/B4/CW	B4
Length:	64ft 6in (19.65m)	66ft 0in (20.11m)
Height:	12ft 9½in (3.89m)	12ft 9½in (3.89m)
Width:	9ft 3in (2.81m)	9ft 3in (2.81m)
Weight:	33 tonnes	32 tonnes
Notes:	Some vehicles have fluorescent lighting	*Some vehicles have disc brakes, and are wired for push-pull operation

Above:

Mk 1 FK. Among the fleet of Mk 1 vehicles constructed from 1951 was a medium sized fleet of FK (First Corridor) vehicles, built either by BR or Metropolitan-Cammell. Many vehicles of the type are now withdrawn and only 33 remain. Each vehicle has seven compartments seating six in each, with a toilet at either end of the coach. FK No 13344 is illustrated from the compartment side. *Michael J. Collins*

2a	2b	2c	2d
13435-13475	13477-13502	13514-13560	13562-13610
1968	1969	1969-70	1971-72
BR Derby	BR Derby	BR Derby	BREL Derby
WR/ER	WR/ER	LMR/WR/ER	LMR/WR/ER
42 first	42 first	42 first	42 first
100mph (161km/h)	100mph (161km/h)	100mph (161km/h)	100mph (161km/h)
30	8*	18*	48
Air*	Air	Air	Air
Dual	Dual	Dual	Electric
4	4	4	5
B4	B4	B4	B4
66ft 0in (20.11m)	66ft 0in (20.11m)	66ft 0in (20.11m)	66ft 0in (20.11m)
12ft 9½in (3.89m)	12ft 9½in (3.89m)	12ft 9½in (3.89m)	12ft 9½in (3.89m)
9ft 3in (2.81m)	9ft 3in (2.81m)	9ft 3in (2.81m)	9ft 3in (2.81m)
33 tonnes	33 tonnes	33 tonnes	35 tonnes
*All built with air brakes but some now converted to vacuum operation. A number of this type are being rebuilt as SK	*Vehicles are being rebuilt into SK stock	*Vehicles are being rebuilt into SK stock	Vehicles 13585-610 require an 800V train supply

Above:
Mk 1 FK. On the corridor side the Mk1 FK has three doors, one at each end and one slightly off-centre. The fleet are mounted on Mk 1, Mk 4 and CW bogies, and brakes of the vacuum or dual types are fitted. Car No 13311 — fitted with vacuum brake, composition brake blocks and CW bogies is illustrated. *Michael J. Collins*

Top:
Mk 1 FK. The identification of the two sides of Mk 1 FK stock is quite easy, as on the corridor side there are three doors while on the compartment side there are only two. Most Mk 1 FKs are painted in blue/grey livery. Car No 13341 which has been painted in the latest InterCity colours. No 13341 is illustrated. *Colin J. Marsden*

Above:
Mk 2 FK. From 1964 the Mk 2 FK fleet emerged. The internal layout followed that established in the Mk 1 build — seven six-seat compartments with a toilet at each end; access was effected by the same door configuration. Internally the compartments have wooden sliding doors, facias and trim. FK No 13389 is illustrated. *Colin J. Marsden*

Top:

Mk 2 FK. View from corridor side of a Mk 2 FK. End connections are basically similar to those found on Mk 1 stock but the toilet water filler points are recessed into the bodywork. The cables hanging either side of the gangway are for train lighting control. The ETH receptacle is seen under the right buffer, and the jumper head to the right of the gangway. Car No 13398 is illustrated. *Colin J. Marsden*

Above:

Mk 2a FK. Mk 2a vehicles are not easily recognised from the Mk 2 design at first glance, but are more identifiable when out of a train formation, as the end doors are of the leaf type coloured green, in preference to the sliding type on earlier builds. When built all Mk 2a vehicles were air-braked, but a number of vacuum conversions have since taken place. *Colin J. Marsden*

Above:
Mk 2b FK. A total of 28 vehicles conforming to Mk 2b specification were built in 1969; all still remain in service although many are now declassified to SK status. Identification of the Mk 2b fleet is not difficult as the vehicles have wrap-around end doors and no centre doors, thus the window dimensions are amended. The vestibule end doors are coloured red. Mk 2b No 13488 is illustrated from the compartment side. *John Tuffs*

Below:
Mk 2c FK. In 1969/70 a fleet of 48 Mk 2c vehicles were built by Derby. Few differences exist between this fleet and the 2b type, except for a revision to the toilet window design and interior decor. Vehicle No 13558 is illustrated. Underframe equipment consists of heat, light, battery and brake units. A number of Mk 2c vehicles have been downgraded to SK status and renumbered into the 19XXX series, the last three digits remaining the same as in their FK days, thus the vehicle illustrated is now numbered 19558.
Colin J. Marsden

Above:
Mk 2d FK. 1971/72 saw the final derivative of the FK emerge when a fleet of 48 Mk 2ds entered service. Recognition of this type is, of course, very easy as the vehicles are air conditioned with no opening passenger compartment windows. Doors are provided at vehicle ends only and accommodation is in seven six-seat compartments. All Mk 2d FK stock has Stones air conditioning equipment and cars 13585-13610 require an 800V train supply. No 13564 is illustrated from the compartment side. *Colin J. Marsden*

Below:
The **FK** fleet are currently allocated to all regions except the ScR, but their numbers are likely to diminish rapidly in the foreseeable future as further converions to SK are effected. A Mk 2a is seen as the first vehicle behind Class 47 No 47122 at Ruscombe in March 1980. *Colin J. Marsden*

Type: **FO** TOPS Code: **AD1**

Mk:	1	2d	2e
Number range:	3021-3150	3170-3216	3221-3275
Introduced:	1953-63	1971-72	1973-74
Built by:	BRC&W/ BR Doncaster/ York/Swindon	BREL Derby	BREL Derby
Regions of allocation:	LMR/WR/ER	LMR/ER	ScR/LMR/WR/ER
Seats:	42 first	42 first	42 first
Speed (max):	90-100mph (145-161km/h)	100mph (161km/h)	100mph (161km/h)
Total number in fleet:	43	47	54
Brake type:	Vacuum/air/dual	Air	Air
Heating type:	Dual	Electric	Electric
ETH index:	3	5	5
Bogie type:	B4/CW	B4	B4
Length:	64ft 6in (19.65m)	66ft 0in (20.11m)	66ft 0in (20.11m)
Height:	12ft 9½in (3.89m)	12ft 9½in (3.89m)	12ft 9½in (3.89m)
Width:	9ft 3in (2.81m)	9ft 3in (2.81m)	9ft 3in (2.81m)
Weight:	32 tonnes	34 tonnes	32 tonnes
Notes:	A number of these vehicles are declassified to SO (AD2)	Nos 3189-3216 require at least 800V train heat supply	Vehicles require at least 800V train heat supply

Above:
Mk 1 FO. The saloon accommodation for first class passengers under the 1951 BR coaching stock scheme was provided by a fleet of FO (First Open) coaches. Seating layout is in the 2+1 style, with a wide off-centre gangway, doors feeding transverse gangways being provided at the ends and towards the centre, dividing the vehicle into three- and four-bay sections. FO No 3066 (now withdrawn) is illustrated from the three-bay end. *Colin J. Marsden*

2f	3a	3b	3a (CO)
3277-3428	11010-11063	11064-11101	11905-11922
1973-75	1975-76	1984-85	1975-76 (orig) as CO 1986
BREL Derby	BREL Derby	BREL Derby	BREL Derby
ScR/LMR/WR/ER	ScR/LMR	LMR	ScR
42 first	48 first	48 first	24 first, 35 second
100mph (161km/h)	110mph (177km/h)	110mph (177km/h)	110mph (177km/h)
107*	52	38	7
Air	Air	Air	Air
Electric	Electric	Electric	Electric
5	6	6	6
B4	BT10†	BT10	BT10
66ft 0in (20.11m)	75ft 0in (22.86m)	75ft 0in (22.86m)	75ft 0in (22.86m)
12ft 9½in (3.89m)	12ft 9in (3.88m)	12ft 9in (3.88m)	12ft 9in (3.88m)
9ft 3in (2.81m)	8ft 11in (2.71m)	8ft 11in (2.71m)	8ft 11in (2.71m)
34 tonnes	32 tonnes	33 tonnes	32 tonnes
*A number of these vehicles are being rebuilt to SO	†11032 has BT15	Cars 11073-11094 are dedicated for 'Pullman' operation	

Above:
Mk 1 FO. Today only 43 Mk 1 FOs remain in service and a number of these are declassified to SO status. Braking systems vary between vacuum, air and dual, but all have dual heating. Several of the later batch of cars numbered from 3107 upwards have fluorescent lighting, aluminium window frames and CW bogies and are owned and operated by Pullman Rail for the charter market. FO No 3113 (No PR125 or PULR99125) is illustrated from the non-toilet side. *Michael J. Collins*

Above:

Mk 2c FO. In 1970 a fleet of 18 Mk 2c FOs were introduced seating 42 in seven six-seat compartments, with toilets at both ends; passenger doors were at vehicle ends only. In 1982-84 the complete fleet were declassified to SO and renumbered 6400-6416. Car No 3164 (now No 6410) is shown. When renumbered as SO stock the only alteration effected, apart from renumbering and painting out of the first class insignia, was the removal of the first class headrests. *Colin J. Marsden*

Below:

Mk 2d FO. From 1971 a fleet of Mk 2d FO vehicles were constructed, differing from the previous builds by incorporating air conditioning, thus having no opening windows except in the four access doors. As in previous FO stock there were two toilets, one at each end. No 3172 is illustrated from the non-toilet side. *Colin J. Marsden*

Above:
Mk 2e FO. A fleet of 55 Mk 2e FO vehicles were built in 1973/74, and are almost identical to the previously described Mk 2d type. From inside, recognition can be achieved as the toilet area is smaller and a luggage stowage area is provided opposite. Both the in and out end vestibule gangways are coloured in fawn. Car No 3251, painted in the latest InterCity colours, is shown from the toilet side. *Michael J. Collins*

Below:
Mk 2f FO. The final variant of the Mk 2 FO design is the 2f of which 164 were built between 1973-75 in three separate lots. The coaches are identical to the previously illustrated type, but interiors are styled with InterCity 70 seats. There is different air conditioning equipment fitted depending on the build, this making a minor difference to underframe modules. Ex-shops No 3372 is shown from the toilet side. *Colin J. Marsden*

Above:
Mk 3a FO. With the development of the IC125 stock in the early 1970s, it was decided to build a fleet of standard locomotive-hauled coaches conforming to the same design and classified Mk 3a. A total of 60 Mk 3a FOs were built and closely resemble the TF vehicle in IC125 formations except for the fitting of buffers and conventional drawgear. There are other small recognition factors such as a narrower first class band on the locomotive-hauled variant and, of course, the omission of the '125' legend on the side. *BR*

Bottom left:
Mk 3b FO. The Mk 3 FO types are laid out in the 2+1 configuration with eight bays, providing a total of 48 seats with a toilet at each end. In 1985 a further 38 Mk 3 FO coaches were constructed designated Mk 3b. The vehicles are basically identical to the Mk 3a except for internal decor and some end alterations including the fitting of integral tail lights and the repositioning of water tank fillers and brake pipe test cocks. Mk 3b No 11100 is illustrated. *Colin J. Marsden*

Above:
Mk 3b FO. In the build of 38 Mk 3b FOs, 22 were constructed for the Pullman market. Again these vehicles have slightly revised interiors and bear the InterCity Pullman legend on the left-hand end. All Pullman FOs are named and this is carried in white on the red waistband, below which is the Pullman motif. Car No 11082 *James Joule* is illustrated. Cars 11083-11094 have telephones installed in one of the luggage stowage areas. *John Tuffs*

Below:
The **FO** breed are allocated to all regions except the SR, and can usually be found in most locomotive-hauled rakes. A Mk 1 FO mounted on CW bogies is seen as the first vehicle behind Class 47 No 47123 near Swindon in early 1980. *Colin J. Marsden*

Type: **RBR** TOPS Code: **AJ4**

Mk: 1
Number range: 1644-1758, 1923-1984
Introduced: 1958-62
Built by: Pressed Steel/BRC&W/BR Swindon
Regions of allocation: LMR/WR/ER
Seats: 23 chairs
Speed (max): 100mph (161km/h)
Total number in fleet: 87
Brake type: Air/dual
Heating type: Dual/electric

ETH index: 2
Bogie type: B4/B5/CW
Length: 64ft 6in (19.65m)
Height: 12ft 9½in (3.89m)
Width: 9ft 3in (2.81m)
Weight: 36-38 tonnes depending on build
Notes: Nos 1644-1762 built as RB, but now all refurbished
Nos 1923-1984 built as RU, but now all rebuilt and refurbished

Below:
Mk 1 RBR. Two distinct vehicle types are classified under the RBR (Restaurant Buffet Refurbished) title, vehicles Nos 1644-1762, and Nos 1923-1984. The first group (1644-1762) were built with 23 loose chairs as RBs. All vehicles are now refurbished. The cooking and counter area is at one end and the seating in four bays at the other. Vehicle No 1665 is seen from the kitchen side, showing the two service door positions. *Colin J. Marsden*

Above right:
Mk 1 RBR. Between the first and second seating bays on both sides of the vehicle is an emergency exit door which can only be opened from the inside. Car No 1693 is illustrated. The four large underframe-mounted boxes carry the gas bottles for cooking. *Michael J. Collins*

Below:
Mk 1 RBR. A RBR viewed from corridor/counter side, with the kitchen and food preparation area at the rear end. There are two toilets in RBR vehicles which are not for public use but for buffet car staff, to conform with legal requirements. Again there are two kitchen access doors, with one public door opposite the bar area. Vehicle No 1672 is shown which is fitted with dual brakes and heat. *Michael J. Collins*

Above:
Mk 1 RBR. This view of vehicle No 1728 shows the corridor/counter side from the saloon end. A number of RBR vehicles are now slowly emerging in the new InterCity colours. Note the underframe steps and side grab-handles by the passenger door. *Colin J. Marsden*

Below:
Mk 1 RBR. The second group of present RBR vehicles were introduced as RU (Restaurant Unclassified). Cars were later rebuilt as RBS (Restaurant Buffet Second), with a counter and 23 fixed chairs, and were again modernised in the 1970s when fluorescent lights were fitted and cars reclassified RBR. If this illustration and the previous one are compared the main detail differences from the saloon end/corridor side can be seen — mainly the omission of the passenger access door. Car No 1923 is illustrated. *Colin J. Marsden*

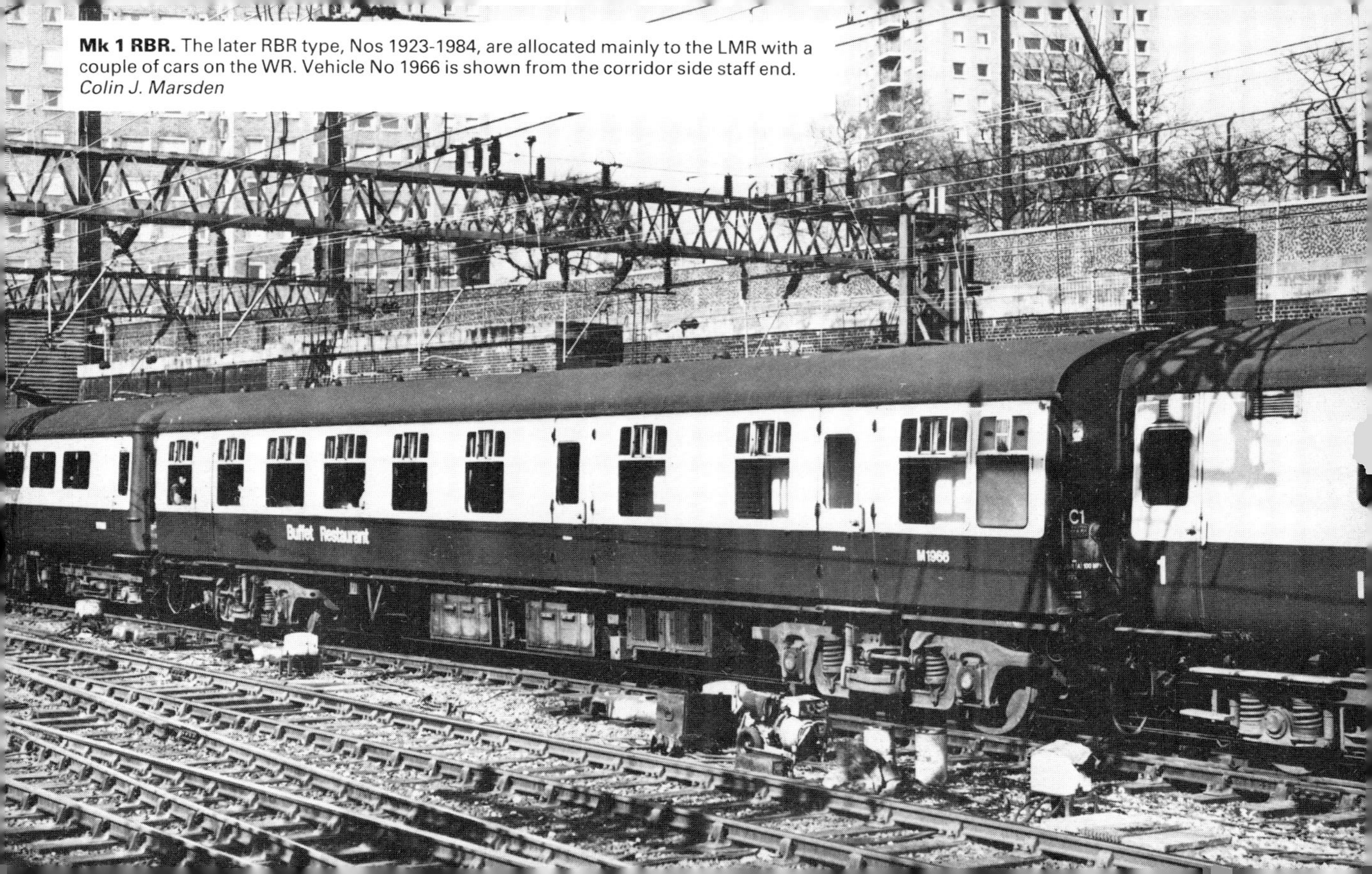

Mk 1 RBR. The later RBR type, Nos 1923-1984, are allocated mainly to the LMR with a couple of cars on the WR. Vehicle No 1966 is shown from the corridor side staff end.
Colin J. Marsden

Type: **RKB** TOPS Code: **AK5**

Mk: 1
Number range: 1508-1567
Introduced: 1960-61
Built by: Cravens
Regions of allocation: ScR/LMR
Seats: None
Speed (max): 100mph (161km/h)
Total number in fleet: 8

Brake type: Vacuum/air
Heating type: Steam/dual
ETH index: 1
Bogie type: B5
Length: 64ft 6in (19.65m)
Height: 12ft 9½in (3.89m)
Width: 9ft 3in (2.81m)
Weight: 41 tonnes

Below:
Mk 1 RKB. For use on longer-distance trains where there are kitchen requirements, a small fleet of RKB (Restaurant Kitchen Buffet) cars are provided. All were built by Cravens Ltd and introduced in 1959. No passenger seating is provided, just a counter with a centre column bar stand. Vehicle No 1526 (now withdrawn) is illustrated from the corridor side, kitchen end. There should be no problem in recognition of RKB stock as the window arrangement is unique to the type. *John Tuffs*

Above right:
Mk 1 RKB. Today only eight RKB vehicles remain in service, and these are normally deployed on long-distance services formed of Mk 1 or Mk 2 stock. In the days when the RKBs were in regular ECML operation prior to the full introduction of IC125s, a 'Deltic'-hauled York-King's Cross train is seen, with an RKB (kitchen side nearest) as the fourth vehicle. *Colin J. Marsden*

Bottom right:
A **Mk 1 RKB** vehicle viewed from the kitchen and counter side. Three staff access doors are provided towards the centre, two of which have additional opening adjacent body panels. An emergency door is provided at one end (farthest in illustration). Most RKBs are now painted in the latest InterCity livery, as displayed here on car No 1528.
Keith Grafton

DW

Kitchen Buffet

Type: **RMB** TOPS Code: **AN2**

Mk: 1
Number range: 1805-1882
Introduced: 1957-62
Built by: BR York/Wolverton
Regions of allocation: ScR/LMR/WR/ER
Seats: 44 second
Speed (max): 100mph (161km/h)
Total number in fleet: 38

Brake type: Vacuum/air/dual
Heating type: Dual
ETH index: 3
Bogie type: B4/B5/CW
Length: 64ft 6in (19.65m)
Height: 12ft 9½in (3.89m)
Width: 9ft 3in (2.81m)
Weight: 38 tonnes

Above:
Mk 1 RMB. The RMB (Restaurant Miniature Buffet) is basically an Open Second coach fitted with a small buffet section towards the centre. Facilities are sparse and only hot drinks and cold snacks can be served satisfactorily. There are two toilets at one end and seating is provided in three bays at one end of the coach and two at the other. Car No 1873 is shown from the counter/storage side from the toilet end. This vehicle and No 1872 were unique in having high-level EMU-compatible jumper facilities. Car Nos 1872, 1873 have now been withdrawn. *Colin J. Marsden*

Left:
Mk 1 RMB. View showing the corridor or counter side from the small seating end, two bays of 2+2 seating. The area directly opposite the counter has two small narrow windows; three passenger access doors are provided on each side. Vehicle No 1881 is illustrated. Underframe equipment visible consists of brake and battery modules. *Colin J. Marsden*

Type: RMBT

TOPS Code: **AB**

Mk: 2d
Number range: 6652-6665
Former number range: 66xx TSO(T) fleet
Introduced: Orig: 1971; as RMBT: 1986
Built by: BREL Derby
Region of allocation: LMR
Seats: —
Speed (max): 100mph (161km/h)
Total number in fleet: 5*
Brake type: Air
Heating type: Electric
ETH index: 5
Bogie type: B4
Length: 66ft 0in (20.11m)
Height: 12ft 9½in (3.89m)
Width: 9ft 3in (2.81m)
Notes: *More vehicles are likely to be converted

Above:
Mk 2d RMBT: To provide buffet facilities for the revised North-South services commencing May 1986 a batch of five RMBT vehicles were converted from existing TSO(T) vehicles. The modification work has dedicated two eight-seat bays to refreshment use. All cars are painted in the latest InterCity colours and operate on the North-South services via Kensington Olympia. Car No 6661 formerly TSO(T) No 6611 is seen under conversion at Ilford. *Brian Morrison*

Type: **RFB** TOPS Code: **AJ1**

Mk: 3a
Number range: 10001-10028
Introduced: 1979-80
Built by: BREL Derby
Region of allocation: LMR
Seats: 17 first
Speed (max): 110mph (177km/h)
Total number in fleet: 28
Brake type: Air
Heating type: Electric
ETH index: 16
Bogie type: BT10
Length: 75ft 0in (22.86m)
Height: 12ft 9in (3.88m)
Width: 8ft 11in (2.71m)
Weight: 40 tonnes
Notes: Classified as RUB until 1984

Below:
Mk 3a RFB. To operate with Mk 3 train formations a fleet of 28 Mk 3a RUB (Restaurant Unclassified Buffet) cars were introduced in 1979/80; each provided 17 unclassified seats in the 2+1 mode, with a cooking and counter area at the opposite end. From 1986, alterations to the LM catering facilities took place and the fleet were reclassified RFB (Restaurant First Buffet). Several of the fleet have been reupholstered and reliveried and now operate on the Pullman diagrams. Car No 10027 is illustrated from the accommodation end, kitchen side. *Colin J. Marsden*

Type: **RSM** TOPS Code: **AJ1**

Mk: 3a
Number range: 10200-10263
Former number range: 40519-405XX, 10001-10028, 110XX-110XX (TRFKs, RFBs and FOs)
Introduced: Originally 1975-79; as RSM: 1986-87
Built by: BREL Derby
Region of allocation: LMR
Seats: 24 second
Speed (max): 110mph (177km/h)
Total number in fleet: 64 (under conversion)
Brake type: Air
Heating type: Electric
ETH index: 14
Bogie type: BT10
Length: 75ft 0in (22.86m)
Height: 12ft 9in (3.88m)
Width: 8ft 11in (2.71m)
Weight: 39.4 tonnes
Notes: This fleet of restaurant cars is under conversion from redundant IC125 TRFKs, Mk 111a RFBs, and FOs. At the time of writing only two vehicles had been converted

Below:
Mk 3a RSM. Major alterations to the Mk 3 catering fleet are now underway. The 12 redundant IC125 TRFKs, all the RFBs and 24 Mk 3a FOs are to be rebuilt into RSMs (Restaurant Second Modular). At the time of writing only two vehicles — Nos 10200/01 rebuilt from TRFKs 40519/20 — were complete. These were initially modified as RFM (Restaurant First Modular), but were later deemed as RSM. Accommodation in the revised vehicle is for 24. Car No 10201 is illustrated.
Colin J. Marsden

Type: **SK** TOPS Code: **AA2**

Mk:	1	2a-c*
Number range:	18250-19207, 24585-24980	194XX and 195XX range
Former number range:	25250-26207	134XX and 135XX FK range
Introduced:	1956-63	1968-70 as FK; 1985-86 as SK
Built by:	Metro-Cammell/Cravens/ BRC&W/BR Derby/ Wolverton	BR/BREL Derby
Regions of allocation:	ScR/LMR/WR/SR/ER	LMR/WR/ER
Seats:	48 or 64 second (depending on build)	42 second
Speed (max):	90-100mph (145-161km/h)	100mph (161km/h)
Total number in fleet:	374	*
Brake type:	Vacuum/air	Air
Heating type:	Steam/dual	Dual
ETH index:	4	4
Bogie type:	B1/B4/CW/single bolster	B4
Length:	64ft 6in (19.65m)	66ft 0in (20.11m)
Height:	12ft 9½in (3.89m)	12ft 9½in (3.89m)
Width:	9ft 3in (2.81m)	9ft 3in (2.81m)
Weight:	33 tonnes	33 tonnes
Notes:		*A number of FK vehicles have been rebuilt into SK stock, and it is understood that the fleet will eventually consist of 53 vehicles.

Above:
Mk 1 SK. The second largest fleet of passenger carrying stock in traffic is of the SK (Second Corridor) type, of which over 400 remain in traffic. The SK fleet have eight compartments fed from a side corridor with two toilets at one end. Three passenger doors are provided on each side. Car No 25798 (now No 18798) is illustrated from the toilet end. This coach is mounted on CW bogies and has vacuum brakes and dual heat. *John Tuffs*

Below:
Mk 1 SK. There are two different internal layouts within the SK fleet — those with and those without armrests — this being reflected in the seating capacity. The majority have armrests and can seat 48, while those without armrests can seat 64. Car No 17582 is shown from the toilet end, compartment side. When this photograph was taken roof destination clips still prevailed. It is usual practice for the four compartments nearest the toilets to be designated for smokers. *Colin J. Marsden*

Above:
Mk 1 SK. All three conventional bogie types — B1, B4 and CW — can be found under the Mk 1 SK stock, the most prolific being the CW and B1, with B4s only being fitted under vehicles 18595-627 which are all allocated to the WR. B4 fitted No 18603 is illustrated. *Michael J. Collins*

Top:
Mk 1 SK. Some minor differences exist on vehicles numbered higher than 18704. These have metal window frames and melamine interior panelling, while many of the earlier cars have their blinds removed. CW-bogied No 18817 is illustrated from the corridor side. *Colin J. Marsden*

Above:
Mk 1 SK. The SK batch Nos 18283-18401 are worthy of special note. These are carried on single-bolster bogies that are similar to the B1 but do not have the normal external coil springs. This difference is noticeable if the two types are compared. Vehicle No 25360 (now No 18360) is illustrated. *Colin J. Marsden*

Top:
Mk 1 SK. Some of the ScR-allocated SK vehicles, were regularly used on the Glasgow-Stranraer service that connected with the Sealink service, and were outshopped in a special Sealink livery of red upper panels and three shades of blue on the lower part, as shown here on vehicle No 18365. These vehicles are now used as charter stock. *Tom Noble*

Above:
During 1985/86 a considerable number of FK vehicles from Mk 2a, 2b and 2c have been declassified to SK standards and renumbered accordingly. The alteration work has mainly consisted of painting out the yellow cantrail band, removing the first class insignia, and removing floor carpets. Car No 19513 (formerly 13513) is shown. While the '1' indicators on the doors have been painted out and the number altered, on this vehicle the yellow band and window legends are still apparent. *Colin J. Marsden*

Type: **SLE/SLEP** TOPS Code: **AS4/AU4**

	SLE	**SLEP**
Mk:	3a	3a
Number range:	10646-10732	10500-10619
Introduced:	1980-84	1981-4
Built by:	BREL Derby	BREL Derby
Regions of allocation:	ScR/LMR/ER	ScR/LMR/ER
Seats:	13 first, or 26 second	12 first, or 24 second
Speed (max):	100mph (161km/h)	100mph (161km/h)
Total number in fleet:	85	117
Brake type:	Air	Air
Heating type:	Electric	Electric
ETH index:	6	7
Bogie type:	BT10	BT10
Length:	75ft 0in (22.86m)	75ft 0in (22.86m)
Height:	12ft 9in (3.88m)	12ft 9in (3.88m)
Width:	8ft 11in (2.71m)	8ft 11in (2.71m)
Weight:	43.5 tonnes	43 tonnes

Below:
Mk 3a SLE. New generation sleeping car stock was ordered in 1980 to replace the rather ancient Mk 1s which dated from the 1950s and were still in service. Two types were ordered — SLE (Sleeper Either Class) and SLEP (Sleeper Either Class with Pantry). This illustration shows a SLE from the compartment or berth side, there being one window for each compartment. If a SLE is designated for second class occupation, each compartment is set for two passengers, giving a vehicle loading of 26 persons. However, if the vehicle is set for first class patrons each compartment is laid out for one only, giving a total coach capacity of just 13. *John Tuffs*

Right:
Mk 3a SLE. The internal fittings of the Mk 3a sleepers are equal to that found in a moderately priced hotel. This view shows a berth set for first class occupancy, with the bed on the left, wash basin towards the centre, and hanging space on the right. Above the window, which has a pull-down blind, is an alarm plunger, shaver socket, and a hammer for breaking the compartment window in an emergency. *BR*

Alarm
In emergency

Above:
Mk 3 SLE. A SLE vehicle viewed from the corridor side, showing the five small windows. Passenger doors are situated one at each corner and are of the standard Mk 3 wrap-around type. Vertical recessed door grips are provided adjacent to each door position. A feature incorporated in the Mk 3 sleeper stock is the provision of integral tail lights, enabling any vehicle to be formed as the last car in a formation, without the provision of a separate oil tail lamp. Car No 10646 is illustrated. *BR*

Below:
Mk 3a SLEP. The Mk 3a SLEP vehicles are almost identical to the SLE fleet except that only 12 passenger compartments are provided. The 13th is replaced by an attendant's pantry for the provision of nighttime drinks and early morning teas; thus the recognition between the two types can be accomplished by observing the single window at either end of the coach. If both are opaque the coach is a SLEP; if one is of standard glass, the coach is a SLE. SLEP No 10580 is illustrated. *Colin J. Marsden*

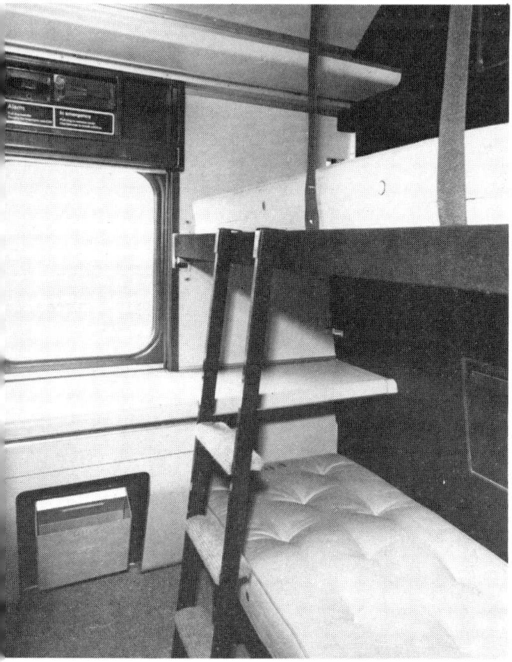

Above:
Mk 3 SLEP. The corridor side of the SLEP is identical to that of the SLE, and for recognition of the two types reference to the number has to be made. In train formation it is usual practice to find one SLEP flanked by two SLEs, therefore one pantry covers three coaches or a maximum of 76 people. Nearly completed No 10531 is shown in Derby Litchurch Lane Works.
Colin J. Marsden

Left:
Mk 3 SLE/SLEP. Interior view showing second class arrangement (two beds). In this illustration the sink top has been lowered to provide a table. *BR*

Mk:	1	1
Number range:	3600-3608	3727-3730, 4358-4372, 4474-4485, 4779-4829
Former number range:	*	—
Introduced:	Orig: 1963; SO: 1981	1954-59
Built by:	BR Swindon	York/BRC&W/BR Derby/Eastleigh
Regions of allocation:	ER	LMR/WR/ER
Seats:	42 second	48 second
Speed (max):	100mph (161km/h)	90-100mph (145-161km/h)
Total number in fleet:	8	54
Brake type:	Air	Vacuum
Heating type:	Dual	Dual
ETH index:	3	4
Bogie type:	CW	B1/B4
Length:	64ft 6in (19.65m)	64ft 6in (19.65m)
Height:	12ft 9½in (3.89m)	12ft 9½in (3.89m)
Width:	9ft 3in (2.81m)	9ft 3in (2.81m)
Weight:	36 tonnes	33 tonnes
Notes:	*Built as FOs in the 31XX range, reclassified as RSO and numbered in the 10XX fleet. Eventually being modified to SO status	

Above:
Mk 1 SO. Although many people classify these vehicles as TSOs, cars Nos 3600-3610, 3727-3730, 4358-4372, 4474-4485 and 4779-4829 are in fact SO (Second Open) vehicles, as their seating is in the 2 + 1 mode. When new, these vehicles were used as second class dining cars but today they operate as SO stock. Cars 3600-3610 were in fact built as FOs and later reclassified RSO (Restaurant Second Open), eventually becoming SOs in 1981/82. SO No 4828 is illustrated. *Colin J. Marsden*

2	2c	2f
5229-5256	6400-6416	6418-6463
—	31XX (FO stock)	32/33XX (FO stock)
1966	Orig: 1970; Rebuilt: 1982	Orig: 1973-75; Rebuilt: 1985-86
BR Derby	BREL Derby	BREL Derby
WR/ER	LMR	LMR/WR
48 second	42 second	42 second
100mph (161km/h)	100mph (161km/h)	100mph (161km/h)
28	17	45
Vacuum/air	Air	Air
Dual	Dual	Electric
4	4	5
B4	B4	B4
66ft 0in (20.11m)	66ft 0in (20.11m)	66ft 0in (20.11m)
12ft 9½in (3.89m)	12ft 9½in (3.89m)	12ft 9½in (3.89m)
9ft 3in (2.81m)	9ft 3in (2.81m)	9ft 3in (2.81m)
32 tonnes	33 tonnes	34 tonnes

Left:
Mk 2 SO. In 1966 Derby Works produced a fleet of 28 SO vehicles with the 2 + 1 layout for restaurant car accommodation. The conventional Mk 2 TSO body shell was used, and external recognition is not easy. SO No 5235 is illustrated from the toilet end.
Michael J. Collins

Below left:
Mk 2c SO. To provide additional second class accommodation, the Mk 2c FO fleet of 17 vehicles were declassified and renumbered as SOs in 1982/84; the internal alterations consisted of removing the headrests and first class carpets. The first of the fleet, No 6400 (formerly 3167), is illustrated. In 1985/86 a further FO/SO conversion took place when 45 Mk 2f FOs were similarly treated. *Michael J. Collins*

Type: **TSO** TOPS Code: **AC2**

Mk:	1	2	2a
Number range:	3738-5067	5070-5228	5257-5433
Introduced:	1953-62	1965-67	1967-68
Built by:	BRC&W/Cravens/ BR Derby/ Doncaster/York/ Ashford/Eastleigh/ Wolverton	BR Derby	BR Derby
Regions of allocation:	ScR/LMR/WR/ SR/ER	ScR/LMR/WR/ER	LMR/WR/ER/ScR
Seats:	64 second	64 second	64 second
Speed (max):	90-100mph (145-161km/h)	100mph (161km/h)	100mph (161km/h)
Total number in fleet:	274	146	164
Brake type:	Vacuum/air/dual	Vacuum/air	Air
Heating type:	Steam/electric/ dual	Dual	Dual
ETH index:	4	4	4
Bogie type:	B1/B4/CW	B4	B4
Length:	64ft 6in (19.65m)	66ft 0in (20.11m)	66ft 0in (20.11m)
Height:	12ft 9½in (3.89m)	12ft 9½in (3.89m)	12ft 9½in (3.89m)
Width:	9ft 3in (2.81m)	9ft 3in (2.81m)	9ft 3in (2.81m)
Weight:	33 tonnes	32 tonnes	32 tonnes
Notes:			

Mk:	2f	3a
Number range:	5908-6184	12004-12168
Introduced:	1973-75	1975-77
Built by:	BREL Derby	BREL Derby
Regions of allocation:	LMR/WR	ScR/LMR
Seats:	64 second	72 second*
Speed (max):	100mph (161km/h)	110mph (177km/h)
Total number in fleet:	207	157
Brake type:	Air	Air
Heating type:	Electric	Electric
ETH index:	5	6
Bogie type:	B4	BT10†
Length:	66ft 0in (20.11m)	75ft 0in (22.86m)
Height:	12ft 9½in (3.89m)	12ft 9in (3.88m)
Width:	9ft 3in (2.81m)	8ft 11in (2.71m)
Weight:	33 tonnes	32 tonnes
Notes:		*Some vehicles seat 80 second in experimental layout †No 12010/62 have BT15 bogies 12139 has BT10a bogies 12140 has BT7 bogies

Right:
Mk 1 TSO. By far the largest fleet in the coaching stock would be the TSO stock (Tourist Second Open) type originally a LNER designation. The Mk 1 TSO, of which there are still 274, can seat 64 in eight bays with two toilets at one end. Car No 4492 is viewed from the toilet end. This coach is mounted on B1 bogies and still retains roof-board clips.
Michael J. Collins

2b	2c	2d	2e
5434-5497	5500-5615	5616-5743	5744-5907
1969	1969-70	1971	1972-73
BR Derby	BR/BREL Derby	BREL Derby	BREL Derby
ScR/LMR/WR/ER	ScR/LMR	ScR/WR/ER	ScR/LMR/WR/ER
62 second	62 second	62 second	64 second
100mph (161km/h)	100mph (161km/h)	100mph (161km/h)	100mph (161km/h)
60	74	106	160
Air	Air	Air	Air
Dual	Dual	Electric	Electric
4	4	5	5
B4	B4	B4	B4
66ft 0in (20.11m)	66ft 0in (20.11m)	66ft 0in (20.11m)	66ft 0in (20.11m)
12ft 9½in (3.89m)	12ft 9½in (3.89m)	12ft 9½in (3.89m)	12ft 9½in (3.89m)
9ft 3in (2.81m)	9ft 3in (2.81m)	9ft 3in (2.81m)	9ft 3in (2.81m)
32 tonnes	32 tonnes	33 tonnes	33 tonnes
			These vehicles require an 800V train heat supply

Top left:
Mk 1 TSO. All Mk 1 TSOs conform to the same basic design, although construction was shared by BR and several private companies. From the toilet end (nearest in picture) there is a full-width vestibule, four seating bays and another vestibule. A number of different permutations of bogie, brake and heat types are to be found. CW-bogied No 4076 is shown, a SR vehicle which has 750V heating. *Colin J. Marsden*

Centre left:
Mk 1 TSO. Mk 1 TSOs numbered above 4831 have some detail differences against earlier examples — all are fitted with fluorescent lighting and a modified design of headrest. Cars numbered higher than 4918 are again slightly revised with metal window frames. In 1984 some vehicles of the later fleet (4918 upwards) were refurbished; this has included new laminate, seat moquette and lighting, but most importantly repainting into the InterCity grey/red scheme. No 5008 is shown. *Colin J. Marsden*

Bottom left:
Mk 2 TSO. After the Mk 1 TSO production the next fleet were of Mk 2 design. The passenger accommodation layout remained as on the Mk 1s and the main difference was in external styling. When built, all vehicles had vacuum brakes and dual heat; however, some were later fitted with air disc-brakes for ScR push-pull operation. Three Mk 2 TSO coaches are seen in this illustration which shows both body sides. *Colin J. Marsden*

Below:
Mk 2a TSO. In 1967/68 Derby Works constructed two batches of Mk 2a TSOs — these were basically the same as the Mk 2s except for the installation of air brakes and dual heat. Passenger accommodation was the same as on the Mk 2 build, except for alterations to the toilets to incorporate foot-pedal flushing. Recognition between the Mk 2 and Mk 2a is not easy, but external identification is helped by the Mk 2a having lime-green folding end doors. Car No 5379 is shown from the saloon end.
Michael J. Collins

Bottom:
Mk 2a TSO. A total of 164 Mk 2a TSOs are currently in service, allocated to all regions except the SR. Underframe equipment consists of light, heat, battery and brake units. Mk 2a TSO No 5381 is illustrated from the vestibule end, painted in the latest Network SouthEast livery. *Chris Wilson*

Top:
Mk 2b TSO. From Mk 2b onwards the Mk 2 fleet have wrap-around doors. With their introduction in 1969 came another important design change when the toilets were placed one at either end and the centre doors were dispensed with. However, these alterations did reduce the accommodation by two. Mk 2b TSO No 5471 is viewed from the toilet side (opaque windows at either end). *Colin J. Marsden*

Above:
Mk 2b TSO. A view showing the vestibule side of the Mk 2b (no toilet window). Underframe equipment visible on car No 5439 is confined to battery and air-brake modules; the vestibule folding end doors on the Mk 2bs are red in colour. The 60 members of the fleet are allocated to the LMR, ER and WR. *Colin J. Marsden*

Top:
Mk 2c TSO. In 1969/70 nearly 100 vehicles conforming to Mk 2c were constructed. They were basically the same as the Mk 2b except for the fitting of lower ceilings with two fluorescent light strips; ducting for air conditioning was also installed but no air conditioning modules were fitted. Recognition between Mk 2b and Mk 2c examples from the outside is not easy, but if the roof can be viewed, only three air vents will be found on the 2c compared to four on the 2b. Car No 5536 is shown from the toilet side.
Colin J. Marsden

Above:
Mk 2c TSO. A minor detail difference concerning the toilet window exists on later Mk 2cs. These are illustrated on No 5581; they are smaller in height than those previously installed, are self-coloured white and have no ventilators above. The Mk 2c fleet, which totals 74 vehicles, is allocated between the LM and Scottish regions.
Michael J. Collins

Above:
Mk 2d TSO. The next major design change in the TSO fleet came in 1971 when the Mk 2d fleet emerged. These were fitted with air conditioning and thus had no opening windows, except those in the doors; two toilets were again provided. Underframe equipment consists of air conditioning, heat, light, battery and brake units. Car No 5650 is shown. The CP after the number indicates the vehicle is allocated to Crown Point, Norwich. *Colin J. Marsden*

Below:
Mk 2e TSO. In 1972 a fleet of 160 Mk 2e TSOs emerged, which were almost identical to the 2ds but with the seating capacity increased by two, giving a total of 64. This was achieved by altering the toilet and vestibule areas, and repositioning the toilets at diagonally opposite corners, thus providing an immediate recognition factor between this fleet and the 2d series. Car 5871 is illustrated. *Colin J. Marsden*

Above:
Mk 2e TSO. If seen in an isolated position the folding end vestibule doors are in a fawn colour, a scheme which continued on the inside of the passenger doors. All vehicles of the build have Stones air conditioning equipment. Car No 5797, painted in the new InterCity colours, is shown. *John Tuffs*

Below:
Mk 2f TSO. The final derivative of the Mk 2 TSO build came in 1973 when the first of 275 Mk 2f vehicles emerged. These were of identical style to the Mk 2e build except for plastic interior panelling and InterCity 70 seats in all but a handful of vehicles (5975/78/84/87/93/94/6005/12/13/18/25/26/36/47/51/53/60-62/6100/39/42-61). Vehicle No 5981 is illustrated. *Colin J. Marsden*

C1
TSO
100 MPH
Air Braked
M 6061

Left:
Mk 2f TSO. The 2f vehicles are fitted with Temperature Ltd air conditioning equipment and thus their underframe modules are slightly different to those under the Mk 2e stock. InterCity-liveried No 6061, a vehicle with earlier designed seats, is illustrated at BREL York. In this livery scheme the white-coloured toilet windows become far more pronounced. *Colin J. Marsden*

Top and above:
Mk 3a TSO. Between 1975/77 164 Mk 3a TSO vehicles were constructed, to the same general design as the IC125 TS stock but, of course, conventional drawgear was fitted. Inside, the cars are the same as IC125 stock with accommodation for 72 in the 2 + 2 layout. There are two toilets, one at each end. All underslung items such as air conditioning, heating and brake units are behind an air-smoothed case, with equipment identified by letters stencilled on the outside. The upper plate shows car No 12071 painted in conventional blue/grey colours, while the lower plate depicts No 12035 in InterCity grey/red colours. Vehicles of the type allocated to Scotland are painted in the ScotRail scheme, where the red band is replaced by a light blue one.
Both: Colin J. Marsden

Type: **TSO(T)** TOPS Code: **AG2**

Mk:	2c	2d
Number range:	6500-6529	6600-6619
Former number range:	55XX, 56XX series TSOs	56XX, 57XX series TSOs
Introduced:	Orig: 1969-70; rebuilt as TSO(T): 1980-81	Orig: 1970; rebuilt as TSO(T): 1980-81
Built by:	BR Derby	BREL Derby
Regions of allocation:	ScR/LMR/WR/ER	ScR/WR
Seats:	54 second	54 second
Speed (max):	100mph (161km/h)	100mph (161km/h)
Total number in fleet:	28	14
Brake type:	Air	Air
Heating type:	Dual	Electric
ETH index:	4	5
Bogie type:	B4	B4
Length:	66ft 0in (20.11m)	66ft 0in (20.11m)
Height:	12ft 9½in (3.89m)	12ft 9½in (3.89m)
Width:	9ft 3in (2.81m)	9ft 3in (2.81m)
Weight:	33 tonnes	33 tonnes
Notes:		Some vehicles now rebuilt as RMBT

Below:
Mk 2c TSO(T). To provide catering facilities on routes where a full refreshment service would prove uneconomic, a fleet of TSO(T) (Tourist Second Open) micro-buffets were converted from members of the Mk 2c and Mk 2d TSO fleets in the early 1980s. The Mk 2c is illustrated here. The conversion work removed one seating bay (eight seats) and one toilet which was converted to a store and staff accommodation. Vehicles are recognisable from TSOs by the red refreshment band above the buffet bay. Vehicle No 6514 is shown from the buffet end. Note the buffet legend on the window.
Colin J. Marsden

Above:
Mk 2c TSO(T). This illustration shows the first of the conversions — No 6500 (formerly 5603) — from the micro-buffet end; however, no legend is on the window and the cantrail markings are a little faded. This vehicle is one of the Mk 2c TSOs that had the smaller opaque toilet windows. The vestibule opposite the counter area is left open for queuing. Facilities offered are very limited with no provision for the heating of water for hot beverages. All hot liquid has to be preheated at stations and kept warm.
Colin J. Marsden

Below:
Mk 2d TSO(T). A batch of 20 Mk 2d TSOs were converted to micro-buffet status in the same way as the Mk 2c vehicles, again losing one seating bay and toilet from public use. Mk 2d TSO(T)s are allocated to the WR and ScR with several of the ScR allocation sporting ScotRail livery. Vehicle No 6602 — one of the WR allocation — is shown.
Colin J. Marsden

Type: **PFB** TOPS Code: **AR1**

Mk: 2
Number range: 586
Introduced: 1966
Built by: BR Derby
Region of allocation: LMR
Seats: 30 first
Speed (max): 100mph (161km/h)
Total number in fleet: 1
Brake type: Vacuum*
Heating type: Electric†
ETH index: 4
Bogie type: B4
Length: 66ft 0in (20.11m)
Height: 12ft 9½in (3.89m)
Width: 9ft 3in (2.81m)
Weight: 35 tonnes
Notes: *To be converted to air braking
†By AC electric locomotives only

Above:
Mk 2 PFB. Although replaced from their 'Manchester Pullman' service in 1985, the LMR have retained some vehicles of the 22 built for special services, and it is likely that the remaining cars will be modernised including the fitting of air brakes. The PFB (Pullman First Brake) fleet now consists of just one vehicle, No 586. The car has guard's facilities with a small luggage area, seating for 30 and one toilet. In common with Pullman practice, inward-opening doors are fitted. *John Tuffs*

Type: **PFK** TOPS Code: **AP1**

Mk: 2
Number range: 504, 506
Introduced: 1966
Built by: BR Derby
Region of allocation: LMR
Seats: 18 first
Speed (max): 100mph (161km/h)
Total number in fleet: 2
Brake type: Vacuum*
Heating type: Electric†
ETH index: 6
Bogie type: B5
Length: 66ft 0in (20.11m)
Height: 12ft 9½in (3.89m)
Width: 9ft 3in (2.81m)
Weight: 40 tonnes
Notes: *To be converted to air braking
†By AC electric locomotives only

Above:
Mk 2 PFK. The fleet of PFKs is now reduced to just two vehicles. These cars provide all cooking facilities for Pullman services. The kitchen area occupies around half the vehicle length with accommodation for 18 at the opposite end. One passenger door and one service door is provided on each side. PFK No 504 is illustrated from the kitchen end, painted in the original Pullman livery of grey body with rail-blue window band.
Colin J. Marsden

Type: **PFP** TOPS Code: **AO1**

Mk: 2
Number range: 546-553
Introduced: 1966
Built by: BR Derby
Region of allocation: LMR
Seats: 36 first
Speed (max): 100mph (161km/h)
Total number in fleet: 7
Brake type: Vacuum*
Heating type: Electric†
ETH index: 6
Bogie type: B4
Length: 66ft 0in (20.11m)
Height: 12ft 9½in (3.89m)
Width: 9ft 3in (2.81m)
Weight: 35 tonnes
Notes: *To be converted to air braking
†By AC electric locomotives only

Above:
Mk 2 PFP. The fleet of PFPs (Pullman First Parlour) are basically FOs with accommodation for 36. There are access doors at each corner of the inward-opening type. Two toilets are provided, one at each end. Currently seven PFP cars remain available for service, and in common with all Mk 2 Pullman vehicles are allocated to Euston Downside. A Class 86-hauled Pullman service formed of PFB, PFP, PFK, PFP, PFP, PFK, PFB is illustrated. *Colin J. Marsden*

Right:
SO. The only non-standard gauge stock operated by BR is on the Vale of Rheidol Railway which comes under the responsibility of the LMR. Three types of vehicle are operated — SO, BCO and BY — the most numerous being the SO of which there are 14 (seven being fully enclosed vehicles and seven semi-opens). This illustration shows enclosed SO No 4146. *Alan Butcher*

Vale of Rheidol Railway Stock

Type:	BY	SO
TOPS Code:	AY2	AX2
Number range:	137	4143-4148, 4994†
Introduced:	1938	1938
Built by:	GWR Swindon	GWR Swindon
Regions of allocation:	LMR	LMR
Seats:	None	56 second
Speed (max):	25mph (40.25km/h)	25mph (40.25km/h)
Total number in fleet:	1	7
Brake type:	Vacuum	Vacuum
Heating type:	None	None
Length:	13ft 0½in (3.96m)	32ft 0in (9.75m)
Width:	6ft 0in (1.87m)	6ft 0in (1.87m)
Notes:		†To be renumbered 4734

Type:	SO	BCO
TOPS Code:	AX2	AY3
Number range:	4149-4151, 4997-5000*†	4993-4996
Introduced:	1938	1938
Built by:	GWR Swindon	GWR Swindon
Regions of allocation:	LMR	LMR
Seats:	48 second	24first, 16 second
Speed (max):	25mph (40.25km/h)	25mph (40.25km/h)
Total number in fleet:	7	2
Brake type:	Vacuum	Vacuum
Heating type:	None	None
Length:	32ft 0in (9.75m)	32ft 0in (9.75m)
Width:	6ft 0in (1.87m)	6ft 0in (1.87m)
Notes:	*Vehicle 4999 is now 'Vista Coach' and seats 30 †4997-5000 to be renumbered 4737-4740	Rebuilt from BSO in 1985

Above:
SO. Only really suitable for good weather operation are the seven semi-open vehicles, as little protection is offered to passengers. SO No 4151 is illustrated. *Alan Butcher*

Below:
BCO. Two BCO (Brake Composite Open) vehicles Nos 4995/4996 are operated. These can accommodate 24 first and 16 second class passengers in two compartments, with a small guard's van/luggage area at one end (nearest in picture). There are three passenger doors on each side. These vehicles were converted from BSOs (Brake Second Open), in 1985. Car No 4996 is illustrated prior to conversion. *BR*

Above:
SO. Open-bodied SO No 4999 has been converted in recent years to a 'Vista Car' (with one side being fully glazed) and having longitudinal first class seating, thus its classification is more correctly FO. The vehicle is seen at Aberystwyth on 5 May 1986.
John Augustson

Below:
A Vale of Rheidol train hauled by No 8 *Llywelyn* is seen approaching Devils Bridge in May 1979. The train is formed SO (enclosed), SO (Open), BSO, and SO (Open).
Michael J. Collins

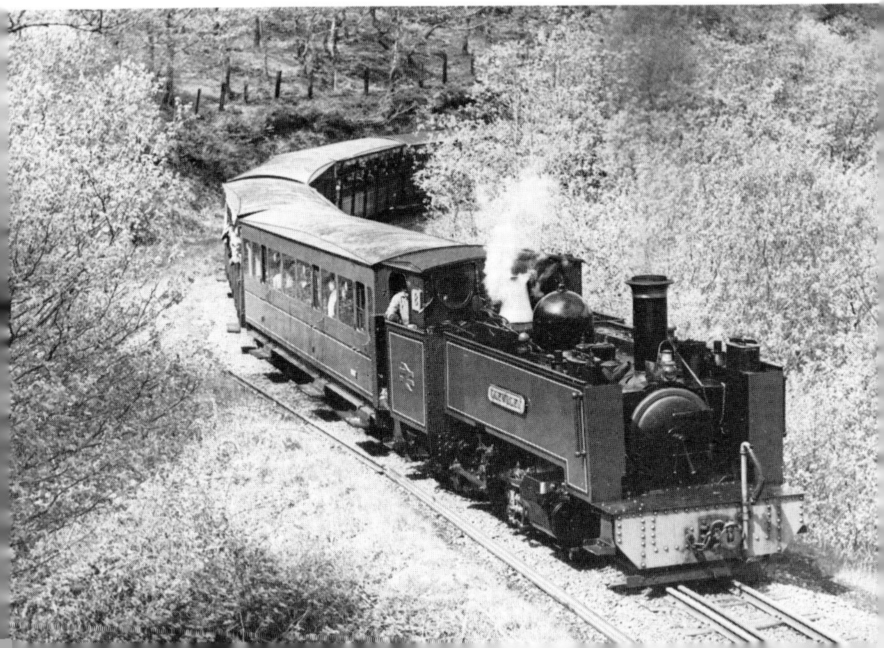

Royal Train Stock

Number:	2900	2901	2902
Former number:			M399
Introduced:	1955	1957	1956
Date vehicle built:	1955	1957	1956
Built by:	BR Wolverton	BR Wolverton	BR Wolverton
Accommodation:	Lounge, Bedroom and Bathroom	Bedroom, Bathroom, Lounge and Office	Kitchen and Dining Saloon
Speed (max):	100mph (161km/h)	100mph (161km/h)	100mph (161km/h)
Brake type:	Air	Air	Air
Heating type:	Electric	Electric	Electric
ETH index:	5	5	3
Bogie type:	B5	B5	B5
Length:	64ft 6in (19.65m)	64ft 6in (19.65m)	64ft 6in (19.65m)
Height:	12ft 9in (3.89m)	12ft 9in (3.89m)	12ft 9in (3.89m)
Width:	9ft 3in (2.81m)	9ft 3in (2.81m)	9ft 3in (2.81m)
Weight:	42 tonnes	42 tonnes	43 tonnes
Notes:			

Number:	2907	2908	2909
Former number:	325	2013	2500
Introduced:	1967	1977	1981
Date vehicle built:	1961	1958	1957
Built by:	BR Swindon	BR Wolverton	BR Doncaster
Accommodation:	Kitchen	Sleeper	Sleeper
Speed (max):	100mph (161km/h)	100mph (161km/h)	100mph (161km/h)
Brake type:	Air	Air	Dual
Heating type:	Electric	Electric	Dual
ETH index:	2	3	3
Bogie type:	B5	B5	B5
Length:	64ft 6in (19.65m)	64ft 6in (19.65m)	64ft 6in (19.65m)
Height:	12ft 9½in (3.89m)	12ft 9½in (3.89m)	12ft 9½in (3.89m)
Width:	9ft 3in (2.81m)	9ft 3in (2.81m)	9ft 3in (2.81m)
Weight:	41 tonnes	40 tonnes	41 tonnes
Notes:		Withdrawn	Withdrawn

Number:	2916-2919
Former number:	40512-40518
Introduced:	1986-87
Date vehicle built:	1976
Built by:	BREL Derby
Accommodation:	Royal Household
Speed (max):	100mph (161km/h)
Brake type:	Air
Heating type:	Electric
ETH index:	6
Bogie type:	BT10
Length:	75ft 0in (22.86m)
Height:	12ft 9in (3.88m)
Width:	8ft 11in (2.71m)

TOPS Code: **AT5**

2903	2904	2905	2906
11001	12011	14105	14112
1977	1977	1977	1977
1972	1972	1969	1969
BREL Derby	BREL Derby	BR Derby	BR Derby
Queen's Saloon	Duke of Edinburgh's Saloon	Brake/Baggage and staff vehicle	Sleeper
100mph (161km/h)	100mph (161km/h)	100mph (161km/h)	100mph (161km/h)
Air	Air	Air	Air
Electric	Electric	Electric	Electric
9	15	5	4
BT10	BT10	B5	B5
75ft 0in (22.86m)	75ft 0in (22.86m)	66ft 0in (20.11m)	66ft 0in (20.11m)
12ft 9in (3.89m)	12ft 9in (3.89m)	12ft 9½in (3.89m)	12ft 9½in (3.89m)
8ft 11in (2.71m)	8ft 11in (2.71m)	9ft 3in (2.81m)	9ft 3in (2.81m)
37 tonnes	37 tonnes	33 tonnes	33 tonnes
		Vehicle also houses generator	

2910	2911	2912	2914-2915
31209	45000	45006	10734-10735*
1941	1982	1942	1985
1941	1920	1942	1985
LMS Wolverton	LMS Wolverton	LMS Wolverton	BREL Derby/ Wolverton
Sleeper, Power Brake	Special Saloon	Special Saloon	Sleeper
70mph (113km/h)	70mph (113km/h)	70mph (113km/h)	100mph (161km/h)
Air	Dual	Air	Air
Electric	Electric	Electric	Electric
3	2	2	6
LMS 6-wheel	B4	LMS 4-wheel	BT10
50ft 0in (15.24m)	57ft 0in (17.37m)	60ft 0in (18.28m)	75ft 0in (22.86m)
12ft 7in (3.83m)	12ft 8in (3.86m)	12ft 4¼in (3.76m)	12ft 9in (3.88m)
9ft 2in (2.79m)	9ft 2in (2.79m)	9ft 3in (2.81m)	8ft 11in (2.71m)
50 tonnes	36 tonnes	42 tonnes	43 tonnes
Spare vehicle	Spare vehicle	Spare vehicle	*Numbers not carried

One of the most famous sets of coaches in the world is the British Royal Train which, when not in use, is housed in special accommodation at Wolverton Works. Coaches are painted in Royal Claret (Purple Lake) livery, and each vehicle has its own special duty to perform within the train. A selection of Royal vehicles are shown but technical data is given for all current stock.

Above:
Often, if a member of the Royal Family is making a tour or staying on the train for several days, much longer formations can be recorded, such as this one at Exeter formed of eight vehicles headed by Class 47 No 47511 *Thames*. This Royal Special photographed on 26 April 1982 was operated from London to Cornwall conveying HRH The Prince of Wales to the Duchy estates. *Colin J. Marsden*

Below:
2900. Royal Saloon 2900 is a lounge/bedroom and bathroom vehicle usually used by members of the Royal Household. The coach has a side corridor, one lounge, four bedrooms and a bathroom. It was purpose-built for royal use at Wolverton in 1955. There are four doors, one at each corner, all of which are inward opening. 2900 was photographed at North Staffs Junction on 23 July 1982. *John Tuffs*

Above:
2901. This saloon was built at Wolverton in 1957 as an office and living accommodation for the Queen's Private Secretary and other members of the Royal Household, but more recently it has been used by HRH The Prince of Wales. The internal layout comprises a side corridor feeding an attendant's compartment, three bedrooms, two bathrooms and an office/lounge. 2901 is illustrated at Exeter on 26 April 1982. *Colin J. Marsden*

Below:
2902. Car 2902 is described as the Royal Dining Car. It was constructed at Wolverton in 1956 as No 399 and was renumbered 2902 in 1977. The layout has a kitchen at one end (far end in picture), and seating accommodation at the other; a small service compartment is also provided. 2902 is illustrated at North Staffs Junction on 23 July 1982. *John Tuffs*

Top:
2903. During the 1977 refurbishment of the Royal Train, two Mk 3 vehicles were converted for Royal use. No 2903 was one of these, being rebuilt from FO No 11001. The coach was considerably rebuilt with twin inward-opening doors on each side. The internal layout consists of a lounge, bedroom and bathroom with a combined bedroom/bathroom for the Queen's dresser. No 2903 is pictured at Stenson Junction in 1982 from the double vestibule door end. *John Tuffs*

Above:
2905. Another 1977 addition to royal stock was 2905 which is a staff couchette/power brake converted from Mk 2b BFK No 14105 of 1969 vintage. In the former brake van there is a 350kW diesel generator, with a small guard's office adjacent. The remainder of the vehicle has luggage accommodation and staff sleeping facilities. 2905 is seen from the staff end. *John Tuffs*

Top:
2906. Converted in 1977 to act as Royal Train escort coach, 2906 was rebuilt from Mk 2b BFK No 14112 and has accommodation for police, day and night accommodation for train staff, guard's facilities and a limited cooking area. 2906 is viewed from the brake end. *Colin J. Marsden*

Above:
2908. A tremendous number of staff — both Royal Household and BR — have to travel and live on the Royal Train when it is in use. For their overnight accommodation two Mk 1 sleeping cars Nos 2013/2500 were taken into royal stock and numbered 2908/9. No 2908 (illustrated) has 11 single berths plus an attendants' compartment with limited cooking facilities, and is viewed from the compartment side. Vehicle Nos 2908, 2909 have now been withdrawn. *Colin J. Marsden*

Type: **BG**

TOPS Code:	NA	NB
Designation:	Standard vehicle	High Security vehicle
Number range:	80517-80975* 84012-84368*	80460-80463
Former number range:	80517-80975 81012-81368	From NDV fleet
Introduced:	1951-63	1957-58
Built by:	Metro-Cammell/ BRC&W/Cravens/ Pressed Steel/ Gloucester/BR Derby/ Eastleigh/York/ Wolverton	Pressed Steel
Regions of allocation:	ScR/LMR/WR/SR/ER	LMR
Load:	10 tonnes	10 tonnes
Speed (max):	90mph (145km/h)	90mph (145km/h)
Total number in fleet:	48	4
Brake type:	Vacuum/dual	Vacuum
Heating type:	Steam	Dual
ETH index:		1
Bogie type:	B1	B1
Length:	57ft 0in (17.37m)	57ft 0in (17.37m)
Height:	12ft 9in (3.88m)	12ft 9in (3.88m)
Width:	9ft 3in (2.81m)	9ft 3in (2.81m)
Weight:	31 tonnes	32 tonnes
Notes:	*Vehicles numbered within given range along with NC and ND types	Further vehicles of this type may be converted

Right:
NA. The standard full brake vehicle, suitable for operation within passenger or vans trains, is the BG (Brake Gangwayed). Today these vehicles are coded under the TOPS NPCCS (Non-passenger Carrying Coaching Stock) system, and there are six groups: NA, NB, NC, ND, NE, and NH. All vehicles are built on a 57ft underframe. The standard vehicle is the NA; these have vacuum brakes, steam heating, and are used for parcels, mail and general brake facilities. Each vehicle has four pairs of doors on each side feeding two luggage vans on either side of a central guard's compartment. All vehicles are mounted on B1 bogies. No 80732 is illustrated. *Colin J. Marsden*

NC	ND	NE	NH
Newspaper vehicle	Dual Heat vehicle	100mph vehicle	110mph vehicle
80503-80946*	80501-80994*	92000-92257*	92000-92257*
84003-84542*	84010-84628*	92308-92383‡	
80503-80946	80501-80994	8XXXX range	8XXXX range
81003-81542	81010-81628		
1951-63	1951-63	1953-63	1953-63
Metro-Cammell/ BRC&W/Cravens/ Pressed Steel/ Gloucester/BR Derby/ Eastleigh/York/ Wolverton	Metro-Cammell/ BRC&W/Cravens/ Pressed Steel/ Gloucester/BR Derby/ Eastleigh/York/ Wolverton	Metro-Cammell/ BRC&W/Cravens/ Pressed Steel/ Gloucester/BR Derby/ Eastleigh/York/ Wolverton	Metro-Cammell/ BRC&W/Cravens/ Pressed Steel/ Gloucester/BR Derby/ Eastleigh/York/ Wolverton
LMR/WR/SR/ER	ScR/LMR/WR/SR/ER	ScR/LMR/WR/ER	ScR/LMR
10 tonnes	10 tonnes	10 tonnes	10 tonnes
90mph (145km/h)	90mph (145km/h)	100mph (161km/h)	110mph (177km/h)
52	360	315	17†
Vacuum	Vacuum/dual	Dual/air	Air
Steam/electric	Dual	Dual	Dual
3	1	1	1
B1	B1	B4/CW	B4 (Special Maintenance)
57ft 0in (17.37m)	57ft 0in (17.37m)	57ft 0in (17.37m)	57ft 0in (17.37m)
12ft 9in (3.88m)	12ft 9in (3.88m)	12ft 9in (3.88m)	12ft 9in (3.88m)
9ft 3in (2.81m)	9ft 3in (2.81m)	9ft 3in (2.81m)	9ft 3in (2.81m)
31 tonnes	31 tonnes	30 tonnes/33 tonnes	30 tonnes
*Vehicles numbered within given range along with NA and ND types	*Vehicles numbered within given range along with NA and NC types	*Vehicles numbered within given range along with NH type	*Vehicles numbered within given range along with NE type †More to be converted

Right:
NC. A number of standard BGs have been converted specifically for the conveyance of newspapers and coded NC. All are fitted with vacuum brakes and are electrically heated. Two variants exist within this type — those retaining guard's accommodation, and those that do not. Most vehicles with guard's accommodation are painted blue/grey, while those with guard's facilities removed are all blue, but this should not be relied upon for total accuracy. Vehicles without guard's facilities have an external parking brake wheel adjacent to the bogie. On some vehicles one or more pairs of doors have been sealed to facilitate packing tables. All are mounted on B1 bogies. Car No 80622 is illustrated. *Colin J. Marsden*

Above:
ND. The ND group is for BG(NA) vehicles fitted with dual heat (steam and electric). Vehicles can be identified from the ends by the presence of both types of connection. Two brake groups exist — vacuum and dual — giving the codes NDV and NDX on vehicle sides. All are mounted on B1 bogies. NDV No 84281 is illustrated.
Colin J. Marsden

Below:
NE. A considerable number of BG vehicles have been rebogied on B4s to enable 100mph operation; these vehicles are coded NE and numbered in the 92XXX range. Internally the vehicles are the same as the NA, except that most are fitted with public address equipment and some have driver/guard communication. A total of 76 ND vehicles are currently under conversion to NE, fitted with CW bogies and authorised for 100mph running; the numbers allocated to this conversion being 92308-92382. At the time of writing most vehicles of the type were in blue/grey livery, but repainting to the latest InterCity colours is expected shortly. Some ScR vehicles are in the ScotRail colour scheme. Car No 92256 is shown. *John Tuffs*

Above:
NH. Selected vehicles — presently 17 of the NE fleet — are now classified NHA; these are identical to NE vehicles but have a maximum permitted speed of 110mph, as a result of the increase in maximum speed on the WCML. All NHAs are allocated to the LMR and painted in the grey/red InterCity colours, vehicle No 92114 being illustrated. *John Tuffs*

Below:
Two **BG** vehicles — one newspaper packing van and a basic vehicle — are seen being shunted at Euston. The BG fleet is allocated to all regions and can be recorded at virtually any location. *Colin J. Marsden*

Type: **B** TOPS Code: **NF**

Number range: S252-S267
Introduced: 1939-1953
Built by: SR/BR Lancing/Ashford
Regions of allocation: SR
Speed (max): 75mph (121km/h)
Total number in fleet: 5
Brake type: Vacuum
Heating type: Steam (if fitted)
Bogie type: SR design
Length: 53ft 8in (16.35m)
Height: 12ft 8in (3.86m)
Width: 9ft 0in (2.74m)
Weight: 26 tonnes
Notes: Scheduled for withdrawal in late 1986

Above:
NFV. A total of 5 SR-designed B vans were still in traffic in mid-1986. These wooden-bodied bogie parcels and newspaper vans are the only restriction 'O' vehicles remaining and are likely to be withdrawn in the near future. Each van has four pairs of doors on either side with a central guard's position. The vans are vacuum-braked, fitted with steam pipes, mounted on SR-design bogies and are able to carry a load of 10 tonnes. *Colin J. Marsden*

Right:
A motley collection of van types have gone to make up this Cornish vans train headed by Class 47 No 47486 near Chasewater on 26 June 1980. From the front the vehicles are: NQV, NPV, NKV, NFV, NKV, NQV, NPV, NKV, NKV and NKV. *Colin J. Marsden*

Type: **Carflat** TOPS Code: **NG**

Number range: 96250-96265
Former number range: B745016-B745048 (fleet)
Introduced: Orig: 1981; to coaching stock fleet: 1985
Region of allocation: LMR
Load: 10 tonnes
Speed (max): 90mph (145km/h)
Total number in fleet: 16
Brake type: Vacuum
Bogie type: B1
Length: 64ft 6in (19.65m)
Width: 9ft 3in (2.81m)
Weight: 23 tonnes
Notes: Flats converted from Mk 1 underframes

Above:
NGV. The latest addition to the coaching stock fleet are 16 Motorail carflats formerly coded FVV; all are allocated to Willesden CS and are operated on Euston-North West services. These flats are former Mk 1 underframes, still retaining buck-eye couplers and rubbing plate, prior to becoming a NGV. FVV No 745031 is illustrated, this vehicle is now 96254. *Colin J. Marsden*

Right:
NK. From 1956 a fleet of nearly 900 GUVs (General Utility Vans) were introduced, constructed by both BR and Pressed Steel Ltd. Today four variants are to be found, all being of the same external appearance. The basic vehicle is classified NK, being fitted with vacuum brake and through steam pipes. Three pairs of doors are fitted on each side together with end doors enabling motor cars to be through loaded. NKV No 86173 (now No 94173) is illustrated. *Colin J. Marsden*

Type: **GUV**

TOPS Code:	NJ	NK
Designation:	Electric wired	Standard vehicle
Number range:	93078-93984	93078-93984
Former number range:	86XXX range	86XXX range
Introduced:	1956-60	1956-60
Built by:	Pressed Steel/ BR York/Doncaster	Pressed Steel/ BR York/Doncaster
Regions of allocation:	LMR/WR/SR/ER	LMR/WR/SR/ER
Load:	14 tonnes	14 tonnes
Speed (max):	90mph (145km/h)	90mph (145km/h)
Total number in fleet:	180	502
Brake type:	Vacuum/dual	Vacuum/dual
Heating type:	Steam pipe/dual-piped	Steam-piped
ETH index:	0	0
Bogie type:	B2	B2
Length:	57ft 0in (17.37m)	57ft 0in (17.37m)
Height:	12ft 4in (3.75m)	12ft 4in (3.75m)
Width:	8ft 6¾in (2.60m)	8ft 6¾in (2.60m)
Weight:	30 tonnes	30 tonnes

TOPS Code:	NL/MV	NX
Designation:	Newspaper vehicle	Motorail vehicle
Number range:	94050-94078	93078-93984
Former number range:	86XXX range	86XXX range
Introduced:	1956-60	1956-60
Built by:	Pressed Steel/ BR York/Doncaster	Pressed Steel/ BR York/Doncaster
Regions of allocation:	LMR/WR/SR/ER	LMR/WR/SR/ER
Load:	14 tonnes	14 tonnes
Speed (max):	90mph (145km/h)	90mph (145km/h)
Total number in fleet:	28	49
Brake type:	Vacuum	Vacuum/dual
Heating type:	Dual	Steam/dual-piped
ETH index:	3	0
Bogie type:	B2	B2/B5
Length:	57ft 0in (17.37m)	57ft 0in (17.37m)
Height:	12ft 4in (3.75m)	12ft 4in (3.75m)
Width:	8ft 6¾in (2.60m)	8ft 6¾in (2.60m)
Weight:	30 tonnes	30 tonnes

Above:
NK. The complete fleet of GUV vehicles is mounted on B2 bogies with the exception of a few Motorail (NX) vehicles that have B5-type bogies. Between the bogies is a handbrake wheel, battery box and brake connections. Some vehicles bear the legend 'Express Parcels' while others are branded for BRUTE traffic and have special securing chains. NKV No 93148 is shown. *Colin J. Marsden*

Below:
NJ. To facilitate electric train heating where GUV vehicles are formed between locomotive and train, a fleet of vehicles have been electrically through-wired and classified NJ. Dual-piped, air-braked No 93749 is illustrated. *Colin J. Marsden*

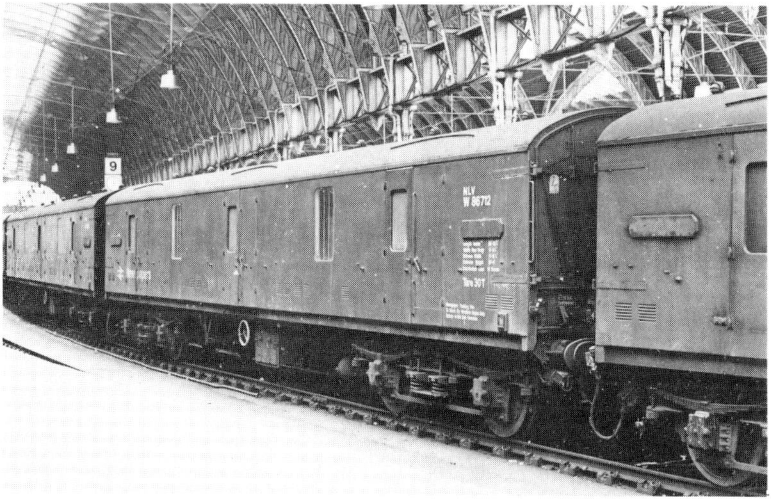

Above:
NLV. The NLV fleet of GUV vehicles are now numbered in the separate series 94050-94078 and are dedicated to newspaper packing, each vehicle bearing the legend 'Newspapers' at the left-hand end of the bodywork. Most newspaper packing vans have hinged internal tables and are dedicated to a set area or route and are often so branded on the outside. Van No 86712 (now 94059) is illustrated. *Colin J. Marsden*

Below:
NX. The NX group within the GUV fleet number 49, are dedicated for Motorail operations and usually bear a suitable bodyside legend. From the outside the cars are the same as the previously illustrated types. The vehicle shown here — No 86949 — carries an additional notice on the side 'To run with Mark 3 sleepers only King's Cross to Aberdeen'. This vehicle is now numbered 93949. *Colin J. Marsden*

Right:
End layout of **GUV** vehicle (NKV): 1. End doors, 2. Door release handle, 3. Drawgear flap, 4. Vacuum pipe, 5. Screw coupling. Most GUV vehicles are painted in all-over blue livery except for a handful which are in blue/grey. *Colin J. Marsden*

Below:
NX: NX vehicles Nos 92308-92383 are being progressively fitted with revised bogies to improve ride, and thus safety. Blue/grey liveried No 93331 is seen sporting the B5 type at Carlisle in July 1986.
John Tuffs

The GUV fleet in their various forms are allocated to all regions, and like the BG fleet can be recorded at virtually any location. This SR vans train is formed of a mixture of GUV and BG vehicles and is seen near St John's (Woking) on 19 April 1984 headed by Class 33 No 33032. *Colin J. Marsden*

Type: **PVG** TOPS Code: **NLX**

Number range: 94001-94034
Former number range: 85501-85534
Introduced: Orig: 1958-59 (as PVG); 1977-78
Built by: Orig: Pressed Steel; as PVG: BREL Wolverton/Doncaster
Regions of allocation: LMR/WR/ER
Load: 14 tonness
Speed (max): 90-100mph (145-161km/h)
Total number in fleet: 35
Brake type: Dual
Heating type: Electric/steam-piped
ETH index: 3
Bogie type: B1/B5
Length: 57ft 0in (17.37m)
Height: 12ft 9in (3.88m)
Width: 8ft 6¾in (2.60m)
Weight: 33-34 tonnes
Notes: Rebuilt from GUV stock

Above:
NLX (PVG). A fleet of vehicles very similar in appearance to the GUV type are these 35 PVG (Packing Van Gangwayed) vehicles now classified NLX. All were converted from GUV stock at Doncaster/Wolverton between 1977-80. The basic difference between the GUV fleet and these PVGs is the fitting of a toilet, washing facilities and newspaper sorting tables with overhead fluorescent lighting. Other alterations made during conversion work were the sealing up of the former end doors, and the fitting of gangways and buck-eye couplings. Vehicles are allocated to the WR, ER and LMR and are usually route branded. Vehicle No 85516 (now 94016) is illustrated with the staff facilities located in the end nearest to the camera. *Colin J. Marsden*

Type: **CCT** TOPS Code: **NO/NP**

Design:	SR	BR
Number range:	S2516	94112-94919
Introduced:	1951-55	1959-61
Built by:	BR Lancing	Earlstown
Regions of allocation:	SR	LMR/WR/SR
Speed (max):	70mph (113km/h)	70mph (113km/h)
Total number in fleet:	1	41
Brake type:	Vacuum	Vacuum
Heating type:	Steam-piped	Steam-piped
Length:	32ft 4½in (9.86m)	37ft 0in (11.27m)
Height:	12ft 8in (3.86m)	12ft 6in (3.81m)
Width:	9ft 0in (2.74m)	8ft 9in (2.66m)
Weight:	14 tonnes	17 tonnes
		Fitted with chains for BRUTE use

Below:
NOV. BR's fleet of parcels vehicles has almost disappeared except for the GUV/BG types. However, a small number of 4-wheeled CCTs (Covered Carriage Trucks) of both SR and BR origin still exist. The SR-designed variant is very similar to the now extinct PMV (NQV), the main difference was the outward-opening end doors on the CCT. All body panelling is wooden and two pairs of doors are provided on each side; Van No 1770 (withdrawn) is illustrated. At the time of writing only one vehicle remains in traffic and was scheduled for withdrawal in late 1986. *Michael J. Collins*

Above:
NOV. Very few of the once 1,000-strong BR CCT variant still exist. The vehicles were constructed of steel and had two pairs of hinged doors on either side, together with end hinged doors to enable through loading. Three body windows (barred inside) were provided on each side, together with operating drop lights in the opening doors. The BR type was fitted with internal lighting for which battery boxes were found on the underframe (opposite side to picture). Van No 94691 is shown. *Colin J. Marsden*

Below:
At the time of writing only 41 examples remain in traffic — mainly allocated to the LMR — with the SR and WR having 10 between them. Class 31 No 31324 stands at Newcastle on 20 May 1982 with the 23.00 vans to Carlisle, with two BR CCTs formed as the first two vehicles. *Colin J. Marsden*

Type: **POS** TOPS Code: **NS**

Mk: 1
Number range: 80300-80395
Introduced: 1959-77
Built by: BR/BREL Wolverton or York
Regions of allocation: LMR/WR/SR/ER
Load: 5-7½ tonnes depending on build
Speed (max): 100mph (161km/h)
Total number in fleet: 95
Brake type: Vacuum/dual
Heating type: Dual
ETH index: 3 or 4
Bogie type: B2/B5
Length: 64ft 6in (19.65m)
Height: 12ft 9½in (3.89m)
Width: 9ft 3in (2.81m)
Weight: 35-39 tonnes depending on build
Notes: Vehicles 80381-95 rebuilds from SK stock

Above:
NS. Many enthusiasts observe the blank face coaches of the Travelling Post Office (TPO) trains during their travels around the railway system. There are three types of TPO vehicle in service, the most numerous being the NS or POS — Post Office Sorting van. These are fitted internally as mobile sorting offices with letter sorting tables and racks along the vehicle sides. Four different batches of NS are in traffic each with distinguishing features. Most NS vehicles have two large sliding doors on one side, and a double-width hinged door on the other. Vehicle No 80311 which is illustrated was built in 1961 at Wolverton, is mounted on B2 bogies and has vacuum brakes and dual heating. *John Tuffs*

Above:
NS. The body-mounted end connections on TPO vehicles include some additions for security purposes. Buffer beam fittings are of the standard type. On the hinged door side there is a posting box into which first class letters may be posted. Car 80353 (illustrated) is one of a fleet of 37 built at York in 1968/69 and is fitted with pressure ventilation equipment, dual brakes and dual heat. *Colin J. Marsden*

Above:
NS. When in operation it is usual for each Post Office Sorting van to be supported by one or more tender vehicles, providing a constant supply of mail. The NS type are allocated to all regions except the ScR. Vehicle No 80359 is illustrated here, this being a 1972 York product fitted with pressure ventilation equipment, fluorescent lighting, B5 bogies, dual brakes and dual heating. All TPO stock carries the Royal Mail legend and coat of arms on the body side. *Colin J. Marsden*

Type: **POT** TOPS Code: **NT**

Mk: 1
Number range: 80400-80439
Introduced: 1959-77
Built by: BR/BREL Wolverton or York
Regions of allocation: LMR/WR/SR/ER
Load: 2-8 tonnes depending on build
Speed (max): 100mph (161km/h)
Total number in fleet: 40
Brake type: Vacuum/dual
Heating type: Dual
ETH index: 3 or 4
Bogie type: B2/B5
Length: 64ft 6in (19.65m)
Height: 12ft 9½in (3.89m)
Width: 9ft 3in (2.81m)
Weight: 34-38 tonnes depending on build
Notes: Vehicles 80400-02 retain net space
Vehicles 80403-14 are rebuilds from BSK
Vehicles 80431-39 are rebuilds from SK

Above:
NT. To provide security stowage for Royal Mail traffic a fleet of NT or POT (Post Office Tender) vehicles are operated. These vehicles have three large sliding doors on either side and the traditional Royal Mail post box towards the centre. Like the NS vehicles, five batches are in service, each having minor detail differences. Vehicle No 80453 is shown, this being a 1977 introduction converted from a SK at Wolverton. The vehicle is pressure ventilated, fitted with fluorescent lighting, is mounted on B5 bogies and is equipped with dual heat and dual brakes. *Colin J. Marsden*

Above:
NT. The three earliest NT vehicles Nos 80400-02 were built with mailbag collection nets. Today this equipment is removed but the space has been left in the body side, giving an identification factor. Cars numbered in the 80403-80414 range introduced in 1959 were converts from BSK stock. The coach illustrated here, No 80419, is a purpose-built Royal Mail stowage vehicle built at York in 1968. Only minor detail differences exist between this and the previously shown vehicle, namely in cantrail ventilation design and end handrails. *Michael J. Collins*

Above:
Most **TPO** services operate with at least three vehicles, but trains such as the West of England or the West Coast TPOs are often seen with 8-10 car rakes. The West of England TPO is seen at Penzance on 1 May 1985 headed by Class 50 No 50005 *Collingwood*. From the front the train is formed NT, NU, NS, NS, NS and NT. *Colin J. Marsden*

Type: **POT (B)** TOPS Code: **NU**

Mk: 1
Number range: 80450-80458
Introduced: 1959-68
Built by: BR Wolverton/York
Regions of allocation: WR/LMR
Load: 8 tonnes
Speed (max): 100mph (161km/h)
Total number in fleet: 9
Brake type: Vacuum/dual
Heating type: Dual
ETH index: 3 or 4
Bogie type: B2/B5
Length: 64ft 6in (19.65m)
Height: 12ft 9½in (3.89m)
Width: 9ft 3in (2.81m)
Weight: 32-35 tonnes depending on build

Below:
To provide guard's facilities on trains formed entirely of TPO vehicles, a fleet of NU or POT-B (Post Office Tender with Brake) vehicles were constructed in 1959, with an additional three in 1968; vehicle No 80455 is illustrated. The sides of the coach have three large sliding doors on each side but towards one end (near in picture) a guard's compartment is located. The earlier built examples Nos 80450-80455 have the window arrangement as shown, whereas the final three only have small cantrail height windows similar to those on the later NT vehicles. *Colin J. Marsden*

Type: **TCV** TOPS Code: **NV**

Number range: 96290-96299
Introduced: 1961-62
Built by: Newton Chambers
Region of allocation: ER
Load: 6 motor cars
Speed (max): 90mph (145km/h)
Total number in fleet: 5
Brake type: Dual
Heating type: Electric-wired, steam-piped
ETH index: 0
Bogie type: Non-standard 6ft 9in (2.06m) wheelbase
Length: 64ft 0in (19.50m)
Height: 12ft 9½in (3.89m)
Width: 9ft 3in (2.81m)
Weight: 32 tonnes

Below:
NVX. This interesting fleet of five NVX or TCV two-tier car vehicles were built in 1961/62 by Newton Chambers to a non-standard design with a fibreglass body and were mounted on unique 6ft 9in wheelbase bogies. End opening doors were provided for loading, and up to six cars (four on upper and two on the lower deck) can be stowed, access to the lower deck between the bogies being via a hydraulic lift. The coaches are painted in the blue/grey scheme and carry the Motorail legend on the left-hand end of each side. All vehicles conform to the UIC Regulations and are fitted with anchor points for transport on train ferries if required. Car No 96294 is illustrated. These vehicles, all of which are allocated to Bounds Green, usually operate between King's Cross and the North East. *Colin J. Marsden*

Type: **SLB/BFB** TOPS Code: **NW**

Number range:	99202-99204	99502-99503
Introduced:	1965-66*	1971
Built by:	Orig: Metro-Cammell	BREL Derby
Regions of allocation:	LMR	M
Load:	10 tonnes	2 or 3 containers
Speed (max):	90mph (145km/h)	80mph (129km/h)
Total number in fleet:	2	2
Brake type:	Dual	Air
Heating type:	Dual	Steam-piped/electric-wired
ETH index:	1	
Bogie type:	B5	B5
Length:	64ft 6in (19.65m)	
Height:	12ft 9½in (3.89m)	
Width:	9ft 3in (2.81m)	
Weight:	30 tonnes	20 tonnes
Notes:	*Rebuilt from BSK stock	

Below:

NW. To provide total security in the movement of bullion and money by rail, a fleet of NW vehicles exist. This group covers two vehicle types — the single-bolster flat used for carrying containers and formerly classified as BFB, and two escort coaches formerly SLB, which have staff accommodation at one end and a large security area at the other. One of the escort vehicles is shown, each of which has two pairs of doors on each side. Sophisticated on-train, and train-to-land communications equipment is carried on these vehicles. The two remaining vehicles are dual braked, dual heated and allocated to the LMR. *Colin J. Marsden*

NW. The bullion flats, (formerly coded BFB) numbered 99500–03, are a single flat, resembling a freightliner vehicle, but having drawgear at each end. Air braking is provided, together with ETH and steam piping; buck-eye couplers and rubbing plates are fitted. Bogies are of the B5 type. Flats 99502/99503 are illustrated. *Michael J. Collins*

Type: **Exhibition Stock** TOPS Code: **NY**
Operated by Ambassador Rail Exhibitions Ltd

Number	Former Number	Introduced	Use
99600	34567	1970	Exhibition
99601	34577	1970	Exhibition
99602	21145	1972	Toilet vehicle
99608	34063	1972	Exhibition
99609	34305	1972	Exhibition
99610	34395	1972	Exhibition
99620	34692	1972	Stores and Brake vehicle
99621	34697	1972	Exhibition
99622	34584	1972	Exhibition
99623	34700	1972	Exhibition
99624	34084	1972	Exhibition
99625	34693	1972	Generator vehicle
99629	25972	1979	Exhibition
99630	34965	1979	Exhibition
99631	1018	1979	Exhibition
99632	4329	1981	Cinema and Lounge vehicle
99633	3827	1981	Cinema and Lounge vehicle
99634	25317	1981	Exhibition
99635	25637	1981	Exhibition
99636	25646	1981	Reception and Toilet vehicle
99637	15854	1981	Exhibition
99638	21221	1981	Exhibition
99641	1036	1981	—
99642	1040	1981	—
99643	4272	1981	Exhibition
99644	3852	1981	Exhibition
99645	1765	1981	Refreshment and Club vehicle
99646	1766	1981	Refreshment and Club vehicle
99647	34629	1981	Exhibition
99648	B787017*	1981	Workshop and Stores vehicle
99649	80502	—	—
99650	80830	—	—
99651	1729	—	—
99652	2000	1985	Sleeper vehicle
99653	2005	1985	Sleeper vehicle
99654	977215	1986	Exhibition
99655	977231	1986	Exhibition
99656	977230	1986	Exhibition
99657	977217	1986	Exhibition
99658	977218	1986	Exhibition
99659	977216	1986	Exhibition
99660	977219	1986	Exhibition
99661	80510	1986	Stores

*Vehicle also carries UIC No 21.70.214.0144-3

Top:
A number of exhibition trains are operated by BR, which are chartered out to some of the large companies and nationalised industries for nationwide promotions. The vehicles are all conversions from Mk 1 locomotive-hauled stock, and when not in operation are usually kept at the exhibition train base at Salisbury. Often windows, and to a lesser extent doors, are altered depending on customers' requirements. Car No 99525 (illustrated) is a generator van rebuilt from a BSK; two exhaust ports can be seen on the roof. *Colin J. Marsden*

Above:
It is usual for exhibition trains to operate in five-, six- or seven-car sets, with reception, exhibition and club facilities. Car No 99632 is a rebuilt Mk 1 TSO that has had its centre doors plated over, toilet removed and venetian blinds fitted. This vehicle is internally laid out as an exhibition area and was converted by BREL Doncaster works in 1981. When this illustration was taken the livery carried was mid-blue, but this varies depending on the requirements of the exhibition organiser. *Colin J. Marsden*

Top:
Rebuilt from SK No 25317, exhibition coach No 99634 is now laid out as an internal exhibition vehicle, all former windows being plated over and now with only four doors. This vehicle is mounted on B1 bogies, has vacuum brakes and is fitted with shore supply power facilities for heating and lighting. Livery when photographed in 1985 was mid-blue with a silver roof. *Colin J. Marsden*

Above:
Painted in orange and white livery, car No 99636 is also a rebuild from a SK (No 25646) but retains some of its original window openings. This car is used as a reception and toilet vehicle and therefore is usually marshalled at the end of a train formation.
Colin J. Marsden

Below:
Also sporting an orange and white scheme is car No 99646, a 1981 conversion by Salisbury of RBR No 1766 to a Club Car (Restaurant facility). Most of the original window openings are retained but a number of internal fittings have been altered. This vehicle is mounted on CW bogies and fitted with vacuum brakes. *Colin J. Marsden*

Bottom:
These former BG vehicles are operated by the exhibition train department — Nos 99649, 99650 and 99661 (formerly Nos 80502, 80830 and 80510); vehicle No 99649 is illustrated. These cars have seen little alteration and are mainly used for stores. Numbers on exhibition stock are only applied on the solebar at the left-hand end. *Colin J. Marsden*

Private Owner Coaching Stock, passed for BR operation

Number	Original Number	Type	Usual location
MCA99050	1	GER Balcony	Steamtown, Carnforth
RSCO99052	45018	LMS Saloon	Steamtown, Carnforth
MCA99053	9004	GWR Saloon	Steamtown, Carnforth
STRM99054	24287	LNER RB	Steamtown, Carnforth
MCA99056	34612	BSK	Steamtown, Carnforth
STRM99064	45005	LMS Royal Saloon	Steamtown, Carnforth
SNG99080	21096	BSK	Steamtown, Carnforth
SGST99103	80972	GWR Saloon	Birmingham RM
SGST99106	9001	GWR Saloon	Birmingham RM
PULR99120	21236	BCK	Bounds Green (BR)
PULR99121	3105	FO	Bounds Green (BR)
PULR99122	3106	FO	Bounds Green (BR)
PULR99123	3109	FO	Bounds Green (BR)
PULR99124	3110	FO	Bounds Green (BR)
PULR99125	3113	FO	Bounds Green (BR)
PULR99126	3116	FO	Bounds Green (BR)
PULR99127	3117	FO	Bounds Green (BR)
PULR99128	3130	FO	Bounds Green (BR)
PULR99129	21272	BCK	Bounds Green (BR)
SVR99240	6913	BCK	Severn Valley Railway
SLOA99335	Car 335	Pullman Kitchen	Bounds Green (BR)
SLOA99347	Car 347	Pullman Kitchen	Bounds Green (BR)
SLOA99348	Car 348	Pullman Kitchen	Bounds Green (BR)
SLOA99349	Car 349	Pullman Kitchen	Bounds Green (BR)
SLOA99350	Car 350	Pullman Kitchen	Bounds Green (BR)
SLOA99351	Car 351	Pullman Kitchen	Bounds Green (BR)
SLOA99352	Car 352	Pullman Kitchen	Bounds Green (BR)
SLOA99353	Car 353	Pullman Kitchen	Bounds Green (BR)
SLOA99354	*Hadrian Bar*	Pullman Second	Bounds Green (BR)
SLOA99355	21249	BCK	Bounds Green (BR)
SLOA99356	21245	BCK	Bounds Green (BR)
SLOA99357	3112	FO	Bounds Green (BR)
PULR99358	3108	FO	Bounds Green (BR)
SLOA99360	2127	SLF	Bounds Green (BR)
VSOE99530	*Perseus*	Pullman First	Stewarts Lane (BR)
VSOE99531	*Phoenix*	Pullman First	Stewarts Lane (BR)
VSOE99532	*Cygnus*	Pullman First	Stewarts Lane (BR)
VSOE99533	70741	LNER BG	Stewarts Lane (BR)
VSOE99534	*Ibis*	Pullman Kitchen	Stewarts Lane (BR)
VSOE99535	*Minerva*	Pullman Kitchen	Stewarts Lane (BR)
VSOE99536	*Zena*	Pullman First	Stewarts Lane (BR)
VSOE99537	*Audrey*	Pullman Kitchen	Stewarts Lane (BR)
VSOE99538	34991	BSK	Stewarts Lane (BR)
VSOE99539	*Ione*	Pullman Kitchen	Stewarts Lane (BR)
VSOE99540	3069	FO	Stewarts Lane (BR)
TRTS99710	18767	SK	Red Bank (BR)
TRTS99711	18857	SK	Red Bank (BR)
TRTS99712	18893	SK	Red Bank (BR)
TRTS99713	19013	SK	Red Bank (BR)
TRTS99714	7187	CK	Red Bank (BR)
TRTS99715	35337	BSK	Red Bank (BR)
MNLP99781	21034	BCK	Bulmers, Hereford
SCR99810	4215	TSO	SRPS Falkirk

SCR99811	1866	RMB	SRPS Falkirk
SCR99812	1513	RKB	SRPS Falkirk
SCR99813	4224	TSO	SRPS Falkirk
SHL99850	35219	BSK	Bulmers, Hereford
DVR99880	4317	TSO	Dart Valley Railway
DVR99881	4046	TSO	Dart Valley Railway
DVR99882	4256	TSO	Dart Valley Railway
NRMY95401	5155	LNWR BFK	NRM York
NRMY95402	*Emerald*	Pullman First	NRM York
NRMY95403	*Eagle*	Pullman Kitchen	NRM York
NRMY99950	35362	BSK	NRM York

Private Owner Operator Codes

NRMY — National Railway Museum, York
MCA — W. McAlpine
RSCO — Resco Railways Ltd
SNG — A4 Locomotive Society
STRM — Steamtown, Carnforth
SGST — Standard Gauge Steam Trust
PULR — Pullman Rail
HLPG — Humberside Locomotive Preservation Society
VSOE — Sea Containers Ltd
SCR — Scottish Railway Preservation Society
SHL — Steam in Hereford Ltd
SVR — Severn Valley Railway
DVR — Dart Valley Railway
MNLP — Merchant Navy Locomotive Preservation Society
TRTS — Train Tours, Manchester
SLOA — Steam Locomotive Operators Association

Above:
A number of former BR and company vehicles now in preservation are passed for main line running. All such vehicles carry a PO number-plate on the left of the solebar prefixed by the owner's code. LMS Saloon No 45018 built in 1927, now carrying the number 41, is BR No RSCO99052 which is usually kept at Steamtown, Carnforth.
Colin J. Marsden

Above left:
Wooden-bodied LNER Restaurant Buffet No 24287 is now preserved by Steamtown, Carnforth, and carries the number STRM99054. All privately owned rolling stock, maintained away from BR premises is subjected to stringent inspection before being permitted onto BR tracks.
Colin J. Marsden

Left:
William McAlpine's 1920-built GER Balcony Saloon No 1 now carries BR operating number MCA99050 and is allocated to Steamtown, Carnforth. This coach is a regular main-line performer and is seen here at Birmingham Moor Street. Vacuum brakes are fitted, as is a through air pipe. *Colin J. Marsden.*

Above:
Twenty-four former BR vehicles are operated by Pullman Rail and housed at Carnforth. However, one set (FO) is usually maintained at Bounds Green. These coaches are made available to the charter market and see regular use. Former BCK No 21272 of 1963 vintage is now PULR 99129 and is mounted on CW bogies and fitted with vacuum brakes and dual heat. Note the PR legend on the body side. *John Tuffs*

Above:
One of the most famous privately owned trains is the Venice Simplon Orient Express (VSOE), which is housed at Stewarts Lane depot, and is used on the regular London-Folkestone run, as well as for nationwide charter work. The 11 VSOE vehicles date from between 1925 and 1957 and represent a beautiful collection of Pullman cars from various manufacturers. 1951-built Pullman Parlour First *Perseus* is illustrated. *John Tuffs*

Below:
Making slow headway towards Folkestone Harbour on 3 June 1982, a nine-vehicle VSOE is seen 'sandwiched' between two Class 73/1 locomotives. The VSOE usually operates twice per week between London and Folkestone, and for the remainder of the week is available to the charter market. *Colin J. Marsden*

Departmental Vehicles

Over the years many hundreds of redundant passenger and NPCCS vehicles have been handed over to the Departmental Board for further use. In most cases this involves the vehicles being completely rebuilt and they often bear little or no resemblance to the original type. This Section illustrates only a handful of these interesting vehicles, and for a more detailed appraisal of the subject reference should be made to the sister volume *Rolling Stock Recognition — Departmental Stock*.

Above:
975036. One of the largest users of former coaching stock vehicles is the Railway Technical Centre at Derby where the R & D, and M&EE technical establishments are housed. No 975036 is formerly Mk 1 BSO No 9234, and is now used as Laboratory 22 in the Tribometer Rail Adhesion Test Train. Considerable external rebuilding has taken place since its traffic days. The livery is red/blue. *Colin J. Marsden*

Below:
975422. Another R&D Section Laboratory vehicle is No 975422 Laboratory 6 *Prometheus*, rebuilt from Mk 1 BCK No 34875 in 1976. This coach has a lower roof section enabling the fitting of a pantograph if required, together with an observation area. When photographed the vehicle was mounted on BT10 bogies. Little alteration has been made to the sides of this coach but the ends now carry special train control jumpers and are painted yellow. The livery is red/blue, and its maximum permitted speed is 150mph. *Colin J. Marsden*

Above:
975002. This vehicle, identified as Laboratory 3, or No 975002, is rebuilt RSO No 1005, and is now the principal laboratory coach for suspension and vehicle development test trains. It carries comprehensive instrumentation including a mini-computer; there are also staff facilities. End observation windows are provided and the coach has a maximum permitted speed of 125mph. Livery is red/blue. *Colin J. Marsden*

Below:
975428. Another vehicle operated by the Research & Development Section is No 975428 or Laboratory 10. This former BSO No 9236 is now operated by the suspension testing department for bogie development. There is a small generator provided in the former brake van and the vehicle is mounted on B4 bogies. Livery is red/blue. *John Tuffs.*

Above:
975136. Although deemed as Laboratory 12 this former 1954-built BSK No 34505 is a stores vehicle for APT development work. Although only having a maximum speed of 100mph this coach has travelled widely with the APT POP development train. The coach is fitted with dual braking, steam heating and is painted in the blue/grey colours.
Colin J. Marsden

Below:
977091. In 1982 Leyland produced an experimental coach body manufactured from bus and coach components and mounted it on the underframe of former Mk 1 CK No 21234. The development of this vehicle was carried out in an attempt to replace the ageing Mk 1 fleet but regrettably the vehicle received adverse publicity and after only a short period of trials disappeared into sidings at Derby, where the vehicle is here seen. Air braking and electric heating were installed. It has since been withdrawn.
Colin J. Marsden

Top:
975814. Not only are older vehicles taken into Departmental service, in 1981 Mk 3 TF No M11000 (latterly W41000) was taken over as RM&EE Test Car No 10. A small auxiliary generator was provided to power test equipment, and an amount of sophisticated recording equipment installed. This coach is painted in red/blue livery and is usually involved when high-speed testing is required. *Colin J. Marsden*

Above:
975948. Still wearing its blue/grey colour scheme, former RU No 1963 is now No 975948, or Test Service Car No 11. The vehicle has been used as an APT Barrier Coach but is now out of use at Derby. All internal fittings have been removed.
Colin J. Marsden

Right:
975672. This former SR BY van No S435 is now classified as QSW and is a match van for SR single- buffer stock. The guard's facilities have been removed but a handbrake is still present. The coach is painted in olive green livery with white lettering and numbering. Brakes are vacuum with through air pipes. *Colin J. Marsden*

QSW
MEs. RSE.
Match Van.
AD8975672

Above:
975666. After the introduction of IC125 stock the movement of individual coaches when 'out of set' became a problem. To overcome this, several barrier vehicles were converted, being suitable to couple between IC125 stock and locomotives. One of the WR barrier vehicles is No 975666 rebuilt from Mk 2a BFK No 14078 in 1978. Livery is all-over blue. *Colin J. Marsden*

Below:
975414. Looking very unlike a passenger coach, No 975414 is a ballast cleaning machine match truck, built on the underframe of SK No 24011, and is thus justified for inclusion in the former passenger stock departmental listings. *Colin J. Marsden*

Above:
975078. Most breakdown train coaches have been rebuilt from redundant passenger stock. No 975078 — a Gateshead-allocated tool van — is rebuilt from BSK No 34828. As can be observed, major modifications have been effected including the panelling over of all original doors and windows, and the forming of two large sliding equipment doors on either side. Livery is yellow. *Colin J. Marsden*

Below:
ADS70133. Some company-built stock still survives in Departmental service such as No ADS70133, the Wimbledon East Staff & Tool coach. This was formerly LSWR 'Ironclad' BTK No S3193. Considerable rebuilding has been effected to this vehicle for its Departmental career such as the fitting of large equipment sliding doors and louvres in the former brake. Livery is yellow. *Colin J. Marsden*

Top:
975453. Part of the Colchester breakdown train unit consists of rerailing equipment van No 975453 which was formerly Mk 1 BSK No 34780. Some external alterations have been made including the removal of the end gangway and the replacement of one pair of luggage van doors with one sliding assembly. Under two of the side windows ventilation grilles have also been installed. Livery is yellow with black chevrons.
Michael J. Collins

Above:
To facilitate the installation and maintenance of overhead live wires a number of purpose-built 'flat top' trains are in service within the Departmental fleet. The 'flat top' coaches are modified Mk 1 vehicles with the roofs specially lowered and flattened to enable men to stand on the train roof whilst working on the overhead equipment. An overhead installation train is illustrated in East Anglia. *Michael J. Collins*

Right:
The Derby-based Departmental coaching stock trains are a regular sight on the Midland main line. Class 31 No 31296 passes Cossington on 18 May 1984 hauling LAB 10 No RDB975428, a Departmental flat wagon No DB945972 and Laboratory 12 No RDB975136. *Colin J. Marsden*

The International Train

During September 1986 BREL launched the International Train: a 10-coach rake of purpose-built exhibition vehicles designed to show potential export customers the type of stock currently available. Over the coming months it is expected that the train will be operated on the LMR on lease to the InterCity Sector, as well as taking part in selective demonstration operations. The 10-vehicle set carries BR running numbers in the private owner fleet series. The train consists of:

99520	Brake First Compartment
99521	First Open, smoking
99522	First Open, non smoking
99523	Catering vehicle, non smoking
99524	Second Open, non smoking, with video equipment
99525	Second Open, smoking, fitted with vending machine
99526	Second Open, smoking
99527	Second Open, non smoking
99528	Second Open, non smoking
99529	Second Open, non smoking

All vehicles are based on the established Mk 3 design, and are mounted on BT10 bogies. The carriages are fitted with swing plug doors; the first InterCity coaches on BR to feature automatic external doors.

Below:
The BREL International Train. *BREL*